T0140367

Studies in Computational Intelligence

Volume 573

Series editor

Janusz Kacprzyk, Polish Academy of Sciences, Warsaw, Poland
e-mail: kacprzyk@ibspan.waw.pl

About this Series

The series "Studies in Computational Intelligence" (SCI) publishes new developments and advances in the various areas of computational intelligence—quickly and with a high quality. The intent is to cover the theory, applications, and design methods of computational intelligence, as embedded in the fields of engineering, computer science, physics and life sciences, as well as the methodologies behind them. The series contains monographs, lecture notes and edited volumes in computational intelligence spanning the areas of neural networks, connectionist systems, genetic algorithms, evolutionary computation, artificial intelligence, cellular automata, self-organizing systems, soft computing, fuzzy systems, and hybrid intelligent systems. Of particular value to both the contributors and the readership are the short publication timeframe and the world-wide distribution, which enable both wide and rapid dissemination of research output.

More information about this series at http://www.springer.com/series/7092

Krassimir T. Atanassov

Index Matrices: Towards an Augmented Matrix Calculus

 Springer

Krassimir T. Atanassov
Department of Bioinformatics
   and Mathematical Modelling
Institute of Biophysics
   and Biomedical Engineering
Bulgarian Academy of Sciences
Sofia
Bulgaria

ISSN 1860-949X
ISSN 1860-9503   (electronic)
ISBN 978-3-319-36504-6
ISBN 978-3-319-10945-9   (eBook)
DOI 10.1007/978-3-319-10945-9

Springer Cham Heidelberg New York Dordrecht London

Printed on acid-free paper

Springer is part of Springer Science+Business Media (www.springer.com)

*The book is devoted to the 145 anniversary
of the Bulgarian Academy of Sciences*

Preface

Let us start with an apparently strange question to the reader. Suppose that we know what a matrix is but we are not familiar with the matrix calculus, or cannot remember how various matrix operations are defined. Suppose that we have two matrices

$$\begin{vmatrix} 1 & 2 & 3 \\ 4 & 5 & 6 \end{vmatrix} \quad \text{and} \quad \begin{vmatrix} 10 & 11 \\ 12 & 13 \\ 14 & 15 \end{vmatrix}$$

and ask quite a natural question on what their sum would be.

Clearly, since the first is a 2×3 matrix and the second is 3×2 matrix, then this question is ill-posed in the well-known classic matrix calculus.

However, apart from that obvious fact, we cannot neglect the fact that the above question can be asked by many people and maybe some reasonable answer could be given.

Some 30 years ago, when I was working on many problems involving various aspects of matrix calculus, mostly in areas that could be described as human centric or human centered, a crucial element had been the human being as a key element of the reasoning or decision-making process. Needless to say that in a vast majority of real-world problems the humans involved are very rarely experts in, for instance, mathematics in general and matrix calculus in particular.

Therefore, I started to think how we could introduce some possibly intuitive and natural changes to the very definitions and elements of matrix calculus, notably the matrices themselves and operations on them to obtain some possibly well-justified results.

So, for instance, in the above ill-posed matrix addition examples in which the matrices involved should be of the same size, that is, 3×3 to avoid truncation of any rows from the second matrix that could result in a substantial loss of information, one could imagine that the following possible extensions could be possible:

$$\begin{vmatrix} 1 & 2 & 3 \\ 4 & 5 & 6 \end{vmatrix} + \begin{vmatrix} 10 & 11 \\ 12 & 13 \\ 14 & 15 \end{vmatrix} \approx \begin{vmatrix} 1 & 2 & 3 \\ 4 & 5 & 6 \\ 0 & 0 & 0 \end{vmatrix} + \begin{vmatrix} 10 & 11 & 0 \\ 12 & 13 & 0 \\ 14 & 15 & 0 \end{vmatrix} = \begin{vmatrix} 11 & 13 & 3 \\ 16 & 18 & 6 \\ 14 & 15 & 0 \end{vmatrix}$$

or

$$\begin{vmatrix} 1 & 2 & 3 \\ 4 & 5 & 6 \end{vmatrix} + \begin{vmatrix} 10 & 11 \\ 12 & 13 \\ 14 & 15 \end{vmatrix} \approx \begin{vmatrix} 1 & 2 & 3 \\ 0 & 0 & 0 \\ 4 & 5 & 6 \end{vmatrix} + \begin{vmatrix} 10 & 11 & 0 \\ 12 & 13 & 0 \\ 14 & 15 & 0 \end{vmatrix} = \begin{vmatrix} 11 & 13 & 3 \\ 12 & 13 & 0 \\ 18 & 20 & 0 \end{vmatrix}$$

or

$$\begin{vmatrix} 1 & 2 & 3 \\ 4 & 5 & 6 \end{vmatrix} + \begin{vmatrix} 10 & 11 \\ 12 & 13 \\ 14 & 15 \end{vmatrix} \approx \begin{vmatrix} 0 & 0 & 0 \\ 1 & 2 & 3 \\ 4 & 5 & 6 \end{vmatrix} + \begin{vmatrix} 10 & 11 & 0 \\ 12 & 13 & 0 \\ 14 & 15 & 0 \end{vmatrix} = \begin{vmatrix} 10 & 11 & 0 \\ 13 & 15 & 3 \\ 18 & 20 & 6 \end{vmatrix}$$

or ...

Clearly, there are nine possible ways or setting such slightly changed, or augmented, formulations of our source ill-posed question which would replace it by a well-posed formulation (question) that is apparently the closest in terms of changes.

Of course, the very essence of the matrix additions as given above is not exactly the same of our source, ill-posed, matrix addition problem, and that is why the symbol "≈" was used which stands for more or less equal, similar, etc. Clearly, this should be meant in a proper way due to the fact that we replace an ill-posed matrix calculus problem by a "similarly looking" well-posed one.

To formalize the above way or reasoning, and to devise some plausible means for handling such problems, I developed first the concept of a so-called "index matrix", and then its corresponding augmented matrix calculus that would make it possible to implement those ideas in a plausible and mathematically correct way.

It turned out later that the new concept, properties, operations, etc., have proved to be extremely useful for solving a multitude of problems in many areas of science and technology in which mathematical modeling, notably based on broadly perceived matrix calculus plays a crucial role.

In particular, I have found that the concept of an index matrix has proved to be very useful in the area of generalized nets, an extension of the Petri nets, I introduced some three decades ago, and which had enjoyed since then a wide popularity as an effective an efficient tool for modeling and solution of a wide array of problems involving systems that can be viewed as being of a discrete event system type. Moreover, I have found a similar importance of the concept of an index matrix in the context of intuitionistic fuzzy sets which I introduced some three decades ago and which, again, had enjoyed since then a booming popularity.

I will therefore present in this book in a comprehensive form the very concept of an index matrix and its related augmented matrix calculus, and will mostly illustrate my exposition with examples related to the generalized nets and intuitionistic fuzzy sets though one should remember that these are just examples of an extremely wide array of possible application areas.

This book contains the basic results of mine over index matrices and some of the open problems concerning them. I will be very glad if the book succeeds in provoking scientific interest and stimulating other fellow researchers to start working in this area.

I am very thankful to my Ph.D. students Velin Andonov, Peter Hadjistoykov, Evgeniy Marinov, Peter Vassilev, and my daughter Vassia Atanassova, who motivated me to prepare the present book and corrected the text and to my coauthors for papers in which the theory of index matrices has been developed: Anthony Shannon (Australia), Eulalia Szmidt and Janusz Kacprzyk (Poland), Veselina Bureva, Deyan Mavrov, Evdokia Sotirova, and Sotir Sotirov (Bulgaria).

I am grateful for the support provided by the projects DFNI-I-01/0006 funded by the National Science Fund, Bulgarian Ministry of Education.

Sofia, Bulgaria, June 2014   Krassimir T. Atanassov

Contents

Chapter 1
Index Matrices: Definitions, Operations, Relations

The concept of Index Matrix (IM) was introduced in 1984 in [2, 3]. During the following 25 years some of its properties were studied, but in general it was used only as an auxiliary tool for describing the transitions of the generalized nets (see [1, 4, 9]), the intuitionistic fuzzy relations and graphs with finite sets of vertices (see [7, 13]) and in some decision making algorithms based on intuitionistic fuzzy estimations (see e.g., [13]).

In the present book, we include the author's results on IMs, published during the last years. They contain the definitions of different types of IMs, as well as the definitions of the operations, relations and operators over IMs.

We start with the basic definition of the concept of an IM with real number elements, following [3, 10]. For brevity, we denote this IM by $\mathcal{R}$-IM.

1.1 Definitions of an Index Matrix with Real Number Elements and (0, 1)-index Matrix

Let $\mathcal{I}$ be a fixed set of indices and $\mathcal{R}$ be the set of real numbers. Let operations $\circ, *: \mathcal{R} \times \mathcal{R} \to \mathcal{R}$ be fixed. For example, they can be the pairs, $\langle \circ, * \rangle \in \{\langle \times, + \rangle, \langle \max, \min \rangle, \langle \min, \max \rangle\}$, or others.

Let the standard sets K and L satisfy the condition: $K, L \subset \mathcal{I}$. Let over these sets, the standard set-theoretical operations be defined. We call "IM with real number elements" ($\mathcal{R}$-IM) the object:

$$[K, L, \{a_{k_i,l_j}\}] \equiv \begin{array}{c|cccc} & l_1 & l_2 & \cdots & l_n \\ \hline k_1 & a_{k_1,l_1} & a_{k_1,l_2} & \cdots & a_{k_1,l_n} \\ k_2 & a_{k_2,l_1} & a_{k_2,l_2} & \cdots & a_{k_2,l_n} \\ \vdots & & & & \\ k_m & a_{k_m,l_1} & a_{k_m,l_2} & \cdots & a_{k_m,l_n} \end{array},$$

© Springer International Publishing Switzerland 2014
K.T. Atanassov, *Index Matrices: Towards an Augmented Matrix Calculus*,
Studies in Computational Intelligence 573, DOI 10.1007/978-3-319-10945-9_1

where

$$K = \{k_1, k_2, \ldots, k_m\} \quad \text{and} \quad L = \{l_1, l_2, \ldots, l_n\},$$

and for $1 \leq i \leq m$, and for $1 \leq j \leq n$: $a_{k_i,l_j} \in \mathcal{R}$.

When set $\mathcal{R}$ is changed with set $\{0, 1\}$, we obtain a particular case of an IM with elements being real numbers, that we denote by $(0, 1)$-IM.

1.2 Operations Over $\mathcal{R}$-IMs and $(0, 1)$-IMs

For the IMs $A = [K, L, \{a_{k_i,l_j}\}]$, $B = [P, Q, \{b_{p_r,q_s}\}]$, operations that are analogous to the usual matrix operations of addition and multiplication are defined, as well as other, specific ones.

Addition

$$A \oplus_{(\circ)} B = [K \cup P, L \cup Q, \{c_{t_u,v_w}\}],$$

where

$$c_{t_u,v_w} = \begin{cases} a_{k_i,l_j}, & \text{if } t_u = k_i \in K \text{ and } v_w = l_j \in L - Q \\ & \text{or } t_u = k_i \in K - P \text{ and } v_w = l_j \in L; \\[2mm] b_{p_r,q_s}, & \text{if } t_u = p_r \in P \text{ and } v_w = q_s \in Q - L \\ & \text{or } t_u = p_r \in P - K \text{ and } v_w = q_s \in Q; \\[2mm] a_{k_i,l_j} \circ b_{p_r,q_s}, & \text{if } t_u = k_i = p_r \in K \cap P \\ & \text{and } v_w = l_j = q_s \in L \cap Q \\[2mm] 0, & \text{otherwise} \end{cases}$$

Of course, here and below, if "$\circ$" is substituted by "$+$", then $a_{k_i,l_j} \circ b_{p_r,q_s} = a_{k_i,l_j} + b_{p_r,q_s}$, while, if "$\circ$" is "max" or min, then $a_{k_i,l_j} \circ b_{p_r,q_s} = \max(a_{k_i,l_j}, b_{p_r,q_s})$ or $a_{k_i,l_j} \circ b_{p_r,q_s} = \min(a_{k_i,l_j}, b_{p_r,q_s})$, respectively.

The geometrical interpretation of operation $\oplus_{(\circ)}$ is

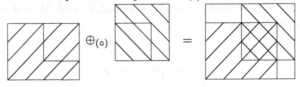

Termwise multiplication

$$A \otimes_{(\circ)} B = [K \cap P, L \cap Q, \{c_{t_u,v_w}\}],$$

where

$$c_{t_u,v_w} = a_{k_i,l_j} \circ b_{p_r,q_s},$$

for $t_u = k_i = p_r \in K \cap P$ and $v_w = l_j = q_s \in L \cap Q$.

The geometrical interpretation of operation $\otimes_{(\circ)}$ is

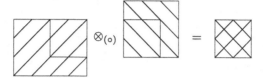

Multiplication

$$A \odot_{(\circ,*)} B = [K \cup (P - L), Q \cup (L - P), \{c_{t_u,v_w}\}],$$

where

$$c_{t_u,v_w} = \begin{cases} a_{k_i,l_j}, & \text{if } t_u = k_i \in K \text{ and } v_w = l_j \in L - P - Q \\[1em] b_{p_r,q_s}, & \text{if } t_u = p_r \in P - L - K \text{ and } v_w = q_s \in Q \\[1em] \underset{l_j = p_r \in L \cap P}{\circ} a_{k_i,l_j} * b_{p_r,q_s}, & \text{if } t_u = k_i \in K \text{ and } v_w = q_s \in Q \\[1em] 0, & \text{otherwise} \end{cases}$$

Obviously, when $\circ$ is substituted by $+$, the symbol $\underset{j}{\circ}$ is substituted by $\underset{j}{\sum}$.

The geometrical interpretation of the multiplication between two ordinary matrices is illustrated as follows,

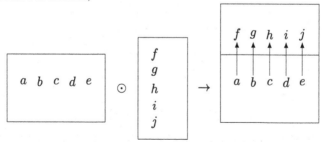

while the geometrical interpretation of the multiplication between two IMs is

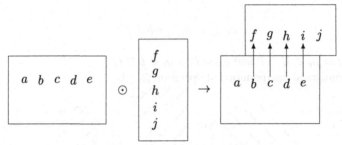

We see immediately that

$$A^{2(\circ,*)} = A \odot_{(\circ,*)} A = [K \cup (K - L), L \cup (L - K), \{c_{t_u,v_w}\}] \quad = [K, L, \{c_{t_u,v_w}\}]$$

where

$$c_{t_u,v_w} = \underset{l_j = p_r \in L \cap P}{\circ} a_{k_i,l_j} * b_{p_r,q_s}$$

and for $n \geq 2$:

$$A^{n+1(\circ,*)} = A^{n(\circ,*)} \odot_{(\circ,*)} A.$$

Structural subtraction

$$A \ominus B = [K - P, L - Q, \{c_{t_u,v_w}\}],$$

where "−" is the set–theoretic difference operation and

$$c_{t_u,v_w} = a_{k_i,l_j}, \text{ for } t_u = k_i \in K - P \text{ and } v_w = l_j \in L - Q.$$

Multiplication with a constant α

$$\alpha A = [K, L, \{\alpha a_{k_i,l_j}\}],$$

where α is a constant.

Termwise subtraction

$$A -_{(+)} B = A \oplus_{(+)} (-1)B.$$

The operation(s) in the sub-index of the operation between IMs, determine(s) the type of operation between the resultant IM-elements.

In the case of (0, 1)-IM, $\circ, * \in \{\min, \max\}$.

It is worth mentioning that for two IMs A and B, such that $K \cap P = L \cap Q = \emptyset$,

$$A \oplus_{(+)} B = A \oplus_{(\times)} B = A \oplus_{(\max)} B = A \oplus_{(\min)} B.$$

For example, if we have the IMs X and Y, and $a, b, c, d, e, p, q, r, s, t, u \in \mathcal{I}$, then

$$
X = \begin{array}{c|ccc} & c & d & e \\ \hline a & 1 & 2 & 3 \\ b & 4 & 5 & 6 \end{array}, \quad
Y = \begin{array}{c|cc} & c & r \\ \hline a & 10 & 11 \\ p & 12 & 13 \\ q & 14 & 15 \end{array},
$$

then

$$
X \oplus_{(+)} Y = \begin{array}{c|cccc} & c & d & e & r \\ \hline a & 11 & 2 & 3 & 11 \\ b & 4 & 5 & 6 & 0 \\ p & 12 & 0 & 0 & 13 \\ q & 14 & 0 & 0 & 15 \end{array}, \quad
X \otimes_{(\times)} Y = \begin{array}{c|c} & c \\ \hline a & 1 \times 10 \end{array} = \begin{array}{c|c} & c \\ \hline a & 10 \end{array},
$$

$$
X \ominus Y = \begin{array}{c|cc} & d & e \\ \hline b & 5 & 6 \end{array}, \quad
X -_{(+)} Y = \begin{array}{c|cccc} & c & d & e & r \\ \hline a & -9 & 2 & 3 & -11 \\ b & 4 & 5 & 6 & 0 \\ p & -12 & 0 & 0 & -13 \\ q & -14 & 0 & 0 & -15 \end{array}.
$$

If IM Z has the form

$$
Z = \begin{array}{c|c} & u \\ \hline c & 10 \\ d & 11 \\ s & 12 \\ t & 13 \end{array},
$$

then

$$
X \odot_{(+,\times)} Z = \begin{array}{c|cc} & e & u \\ \hline a & 3 & 1 \times 10 + 2 \times 11 \\ b & 6 & 4 \times 10 + 5 \times 11 \\ s & 0 & 12 \\ t & 0 & 13 \end{array} = \begin{array}{c|cc} & e & u \\ \hline a & 3 & 32 \\ b & 6 & 95 \\ s & 0 & 12 \\ t & 0 & 13 \end{array}.
$$

For the case of (0, 1)-IM, operation (1.6) has the form

$$
A -_{(\circ)} B = [K \cup P, L \cup Q, \{c_{t_u, v_w}\}],
$$

where

$$
c_{t_u,v_w} = \begin{cases} a_{k_i,l_j}, & \text{if } t_u = k_i \in K \text{ and } v_w = l_j \in L - Q \\ & \text{or } t_u = k_i \in K - P \text{ and } v_w = l_j \in L; \\ \circ(a_{k_i,l_j}, b_{p_r,q_s}), & \text{if } t_u = k_i = p_r \in K \cap P \\ & \text{and } v_w = l_j = q_s \in L \cap Q \\ 0, & \text{otherwise} \end{cases},
$$

and $\circ(a_{k_i,l_j}, b_{p_r,q_s}) = \max(0, a_{k_i,l_j} - b_{p_r,q_s})$.

Because operation $\odot_{(\circ,*)}$ is one of the most complex, we illustrate it with two examples. First, let us have the IM

$$
A = \begin{array}{c|ccc}
 & d & e & f \\ \hline
a & \alpha & \beta & \gamma \\
b & \delta & \varepsilon & \zeta \\
c & \eta & \theta & \iota \\
d & \kappa & \lambda & \mu
\end{array},
$$

where $\alpha, \beta, \ldots, \mu \in \{0, 1\}$. Then

$$
A^2 = A \odot_{(\max,\min)} A = \begin{array}{c|ccc}
 & d & e & f \\ \hline
a & \alpha & \beta & \gamma \\
b & \delta & \varepsilon & \zeta \\
c & \eta & \theta & \iota \\
d & \kappa & \lambda & \mu
\end{array} \odot_{(\max,\min)} \begin{array}{c|ccc}
 & d & e & f \\ \hline
a & \alpha & \beta & \gamma \\
b & \delta & \varepsilon & \zeta \\
c & \eta & \theta & \iota \\
d & \kappa & \lambda & \mu
\end{array}
$$

$$
= \begin{array}{c|ccc}
 & d & e & f \\ \hline
a & \min(\alpha, \kappa) & \min(\alpha, \lambda) & \min(\alpha, \mu) \\
b & \min(\delta, \kappa) & \min(\delta, \lambda) & \min(\delta, \mu) \\
c & \min(\eta, \kappa) & \min(\eta, \lambda) & \min(\eta, \mu) \\
d & \min(\kappa, \kappa) & \min(\kappa, \lambda) & \min(\kappa, \mu)
\end{array}.
$$

Indeed, the pair of indices (a, d) in the resultant matrix is met in the two IMs (that coincide) and therefore, we cannot use the first two parts of formula for multiplication. In this case, we must use the third part of the same formula. Because $\{a, b, c, d\} \cap \{d, e, f\} = \{d\}$, the third operation will be reduced to the form $\min(x, y)$. Therefore, the element of the resultant IM, determined by the pair of indices (a, d) is exactly $\min(\alpha, \kappa)$. All other elements of the resultant IM are obtained in similar ways. Now, let us have the IM

$$
B = \begin{array}{c|ccc}
 & d & e & f \\ \hline
a & \alpha & \beta & \gamma \\
b & \delta & \varepsilon & \zeta \\
e & \eta & \theta & \iota \\
d & \kappa & \lambda & \mu
\end{array}.
$$

where $\alpha, \beta, \ldots, \mu \in \{0, 1\}$. Then we obtain sequentially[1]

$$
B^2 = B \odot_{(\min,\max)} B =
\begin{array}{c|ccc}
 & d & e & f \\
\hline
a & \alpha & \beta & \gamma \\
b & \delta & \varepsilon & \zeta \\
e & \eta & \theta & \iota \\
d & \kappa & \lambda & \mu
\end{array}
\odot_{(\min,\max)}
\begin{array}{c|ccc}
 & d & e & f \\
\hline
a & \alpha & \beta & \gamma \\
b & \delta & \varepsilon & \zeta \\
e & \eta & \theta & \iota \\
d & \kappa & \lambda & \mu
\end{array}
$$

$$
=
\begin{array}{c|ccc}
 & d & e & \ldots \\
\hline
a & \min(\max(\alpha, \kappa)), \max(\beta, \eta)) & \min(\max(\alpha, \lambda), \max(\beta, \theta)) & \ldots \\
b & \min(\max(\delta, \kappa)), \max(\varepsilon, \eta)) & \min(\max(\delta, \lambda), \max(\varepsilon, \theta)) & \ldots \\
e & \min(\max(\eta, \kappa)), \max(\theta, \eta)) & \min(\max(\eta, \lambda), \theta) & \ldots \\
d & \min(\kappa, \max(\lambda, \eta)) & \min(\max(\kappa, \lambda), \max(\varepsilon, \theta)) & \ldots
\end{array}
$$

$$
\begin{array}{c|cc}
 & \ldots & f \\
\hline
a & \ldots & \min(\max(\alpha, \mu), \max(\beta, \iota)) \\
b & \ldots & \min(\max(\delta, \mu), \max(\varepsilon, \iota)) \, . \\
e & \ldots & \min(\max(\eta, \mu), \max(\theta, \iota)) \\
d & \ldots & \min(\max(\kappa, \mu), \max(\lambda, \iota))
\end{array}
$$

Now, we can directly see that when $K = P = \{1, 2, \ldots, m\}$ and $L = Q = \{1, 2, \ldots, n\}$ we obtain the definitions for standard matrix operations.

In the IM case, we can use different symbols as indices of the rows and columns and they, as we saw above, give us additional information and possibilities for description.

Let $\mathcal{IM}_{\mathcal{R}}$ be the set of all $\mathcal{R}$-IMs and let

$$I_{\emptyset} = [\emptyset, \emptyset, \perp],$$

where symbol "$\perp$" denotes the lack of IM-elements. Let, as above, $\circ, * \in \{+, \times, \max, \min\}$ and $(\circ, *) \in \{(+, \times), (\max, \min), (\min, \max)\}$.

The following assertions for the IM are discussed in [3].

Theorem 1 (a) $\langle \mathcal{IM}_{\mathcal{R}}, \oplus_{\circ} \rangle$ *is a commutative semigroup,*
(b) $\langle \mathcal{IM}_{\mathcal{R}}, \otimes_{\circ} \rangle$ *is a commutative semigroup,*
(c) $\langle \mathcal{IM}_{\mathcal{R}}, \odot_{\circ,+} \rangle$ *is a semigroup,*
(d) $\langle \mathcal{IM}_{\mathcal{R}}, \oplus_{\circ}, I_{\emptyset} \rangle$ *is a commutative monoid.*

Let the IM A be given and let $k_0 \notin K$ and $l_0 \notin L$ be two indices. Now, following the paper of E. Sotirova, V. Bureva and the author [24], we introduce the following eight aggregation operations over it:

---

[1] Here, in the IM we use symbol "…" in the end of the rows of the first IM and in the beginning of the second IM to denote that the IM is divided into two parts, because of a lack of place.

Max-row-aggregation

$$\rho_{\max}(A, k_0) = \frac{\begin{array}{c|ccccc} & l_1 & l_2 & \ldots & l_n \\ \hline k_0 & \max\limits_{1 \leq i \leq m}(a_{k_i, l_1}) & \max\limits_{1 \leq i \leq m}(a_{k_i, l_2}) & \ldots & \max\limits_{1 \leq i \leq m}(a_{k_i, l_n}) \end{array}}{},$$

Min-row-aggregation

$$\rho_{min}(A, k_0) = \frac{\begin{array}{c|ccccc} & l_1 & l_2 & \ldots & l_n \\ \hline k_0 & \min\limits_{1 \leq i \leq m}(a_{k_i, l_1}) & \min\limits_{1 \leq i \leq m}(a_{k_i, l_2}) & \ldots & \min\limits_{1 \leq i \leq m}(a_{k_i, l_n}) \end{array}}{},$$

Sum-row-aggregation

$$\rho_{\mathrm{sum}}(A, k_0) = \frac{\begin{array}{c|ccccc} & l_1 & l_2 & \ldots & l_n \\ \hline k_0 & \sum\limits_{i=1}^{m} a_{k_i, l_1} & \sum\limits_{i=1}^{m} a_{k_i, l_2} & \ldots & \sum\limits_{i=1}^{m} a_{k_i, l_n} \end{array}}{},$$

Average-row-aggregation

$$\rho_{\mathrm{ave}}(A, k_0) = \frac{\begin{array}{c|ccccc} & l_1 & l_2 & \ldots & l_n \\ \hline k_0 & \frac{1}{m}\sum\limits_{i=1}^{m} a_{k_i, l_1} & \frac{1}{m}\sum\limits_{i=1}^{m} a_{k_i, l_2} & \ldots & \frac{1}{m}\sum\limits_{i=1}^{m} a_{k_i, l_n} \end{array}}{},$$

Max-column-aggregation

$$\sigma_{\max}(A, l_0) = \frac{\begin{array}{c|c} & l_0 \\ \hline k_1 & \max\limits_{1 \leq j \leq n} a_{k_1, l_j} \\ \vdots & \vdots \\ k_m & \max\limits_{1 \leq j \leq n} a_{k_m, l_j} \end{array}}{},$$

Min-column-aggregation

$$\sigma_{min}(A, l_0) = \frac{\begin{array}{c|c} & l_0 \\ \hline k_1 & \min\limits_{1 \leq j \leq n} a_{k_1, l_j} \\ \vdots & \vdots \\ k_m & \min\limits_{1 \leq j \leq n} a_{k_m, l_j} \end{array}}{},$$

Sum-column-aggregation

$$\sigma_{\text{sum}}(A, l_0) = \begin{array}{c|c} & l_0 \\ \hline k_1 & \sum\limits_{j=1}^{n} a_{k_1, l_j} \\ \vdots & \vdots \\ k_m & \sum\limits_{j=1}^{n} a_{k_m, l_j} \end{array} \quad ,$$

Average-column-aggregation

$$\sigma_{\text{ave}}(A, l_0) = \begin{array}{c|c} & l_0 \\ \hline k_1 & \frac{1}{n}\sum\limits_{j=1}^{n} a_{k_1, l_j} \\ \vdots & \vdots \\ k_m & \frac{1}{n}\sum\limits_{j=1}^{n} a_{k_m, l_j} \end{array} \quad .$$

We can see immediately that for every IM A, for every pair of indices i and j and for every $\circ \in \{\max, \min, \text{sum}, \text{ave}\}$:

(1) $\rho_\circ(\rho_\circ(A, j), i) = \rho_\circ(A, i)$,

(2) $\sigma_\circ(\sigma_\circ(A, i), j) = \sigma_\circ(A, j)$,

(3) $\rho_\circ(\sigma_\circ(A, j), i) = \sigma_\circ(\rho_\circ(A, i), j)$.

In the case of (0, 1)-IMs, only operations Max-row-, Min-row-, Max-column- and Min-column aggregation are possible.

1.3 Relations Over $\mathcal{R}$-IM and (0, 1)-IM

Let the two IMs $A = [K, L, \{a_{k,l}\}]$ and $B = [P, Q, \{b_{p,q}\}]$ be given. We shall introduce the following (new) definitions where $\subset$ and $\subseteq$ denote the relations *"strong inclusion"* and *"weak inclusion"*.

The strict relation "inclusion about dimension" is

$$A \subset_d B \text{ iff } (((K \subset P)\&(L \subset Q)) \vee ((K \subseteq P)\&(L \subset Q))$$

$$\vee((K \subset P)\&(L \subseteq Q))) \& (\forall k \in K)(\forall l \in L)(a_{k,l} = b_{k,l}).$$

The non-strict relation "inclusion about dimension" is

$$A \subseteq_d B \text{ iff } (K \subseteq P) \& (L \subseteq Q) \& (\forall k \in K)(\forall l \in L)(a_{k,l} = b_{k,l}).$$

The strict relation "inclusion about value" is

$$A \subset_v B \text{ iff } (K = P) \& (L = Q) \& (\forall k \in K)(\forall l \in L)(a_{k,l} < b_{k,l}).$$

The non-strict relation "inclusion about value" is

$$A \subseteq_v B \text{ iff } (K = P) \& (L = Q) \& (\forall k \in K)(\forall l \in L)(a_{k,l} \leq b_{k,l}).$$

The strict relation "inclusion" is

$$A \subset_* B \text{ iff } (((K \subset P) \& (L \subset Q)) \vee ((K \subseteq P) \& (L \subset Q))$$

$$\vee ((K \subset P) \& (L \subseteq Q))) \& (\forall k \in K)(\forall l \in L)(a_{k,l} < b_{k,l}).$$

The non-strict relation "inclusion" is

$$A \subseteq_* B \text{ iff } (K \subseteq P) \& (L \subseteq Q) \& (\forall k \in K)(\forall l \in L)(a_{k,l} \leq b_{k,l}).$$

It can be directly seen that for every two IMs A and B,

- if $A \subset_d B$, then $A \subseteq_d B$;
- if $A \subset_v B$, then $A \subseteq_v B$;
- if $A \subset B$, $A \subseteq_d B$, or $A \subseteq_v B$, then $A \subseteq B$;
- if $A \subset_d B$ or $A \subset_v B$, then $A \subseteq B$.

Similar properties are valid for the relations, discussed in the next chapters and by this reasons, they are not mentioned.

1.4 Index Matrices with Elements Logical Variables, Propositions or Predicates

When we choose to work with matrices, elements of which are logical variables, propositions or predicates—let us call these IM "Logical IMs (L-IMs)", the form of the IM from Sect. 1.1 is kept, but there are differences in the forms of the operations and relations.

The operations from Sect. 1.2 now obtain the following forms.

$A \oplus_{(\circ)} B = [K \cup P, L \cup Q, \{c_{t_u, v_w}\}]$, where

$$c_{t_u,v_w} = \begin{cases} a_{k_i,l_j}, & \text{if } t_u = k_i \in K \text{ and } v_w = l_j \in L - Q \\ & \text{or } t_u = k_i \in K - P \text{ and } v_w = l_j \in L; \\[2ex] b_{p_r,q_s}, & \text{if } t_u = p_r \in P \text{ and } v_w = q_s \in Q - L \\ & \text{or } t_u = p_r \in P - K \text{ and } v_w = q_s \in Q; \\[2ex] a_{k_i,l_j} \circ b_{p_r,q_s}, & \text{if } t_u = k_i = p_r \in K \cap P \\ & \text{and } v_w = l_j = q_s \in L \cap Q \\[2ex] false, & \text{otherwise} \end{cases}$$,

where here and below $\circ \in \{\wedge, \vee, \rightarrow, \equiv\}$.
$A \otimes_{(\circ)} B = [K \cap P, L \cap Q, \{c_{t_u,v_w}\}]$, where

$$c_{t_u,v_w} = a_{k_i,l_j} \circ b_{p_r,q_s}, \text{ for } t_u = k_i = p_r \in K \cap P \text{ and }$$
$$v_w = l_j = q_s \in L \cap Q;$$

$A \odot_{(\circ,*)} B = [K \cup (P - L), Q \cup (L - P), \{c_{t_u,v_w}\}]$, where

$$c_{t_u,v_w} = \begin{cases} a_{k_i,l_j}, & \text{if } t_u = k_i \in K \\ & \text{and } v_w = l_j \in L - P - Q \\[2ex] b_{p_r,q_s}, & \text{if } t_u = p_r \in P - L - K \\ & \text{and } v_w = q_s \in Q \\[2ex] \underset{l_j = p_r \in L \cap P}{\circ} (a_{k_i,l_j} * b_{p_r,q_s}), & \text{if } t_u = k_i = p_r \in K \text{ and } v_w = q_s \in Q \\[2ex] false, & \text{otherwise} \end{cases}$$,

where $(\circ, *) \in \{(\vee, \wedge), (\wedge, \vee)\}$.

Operation (1.4) from Sect. 1.2 preserves its form, while operations (1.5) and (1.6) from the same section are impossible in general. They are possible only in the case, when α is the operation negation. In this case, operations (1.5) and (1.6) obtain the forms

$$\neg A = [K, L, \{\neg a_{k_i,l_j}\}],$$

$$A -_{(\circ)} B = [K \cup P, L \cup Q, \{c_{t_u,v_w}\}],$$

where

$$c_{t_u,v_w} = \begin{cases} a_{k_i,l_j}, & \text{if } t_u = k_i \in K \text{ and } v_w = l_j \in L - Q \\ & \text{or } t_u = k_i \in K - P \text{ and } v_w = l_j \in L; \\[2mm] \neg b_{p_r,q_s}, & \text{if } t_u = p_r \in P \text{ and } v_w = q_s \in Q - L \\ & \text{or } t_u = p_r \in P - K \text{ and } v_w = q_s \in Q; \\[2mm] a_{k_i,l_j} \circ \neg b_{p_r,q_s}, & \text{if } t_u = k_i = p_r \in K \cap P \\ & \text{and } v_w = l_j = q_s \in L \cap Q \\[2mm] false, & \text{otherwise} \end{cases}$$

and "$\neg$" means operation "negation". This notation is used, because of the variety of different possible forms of the operation "negation", as discussed in Sect. 2.1.

Only operations Max-row-, Min-row-, Max-column- and Min-column aggregation from Sect. 1.2 exist for L-IMs, but in the following forms.

$\vee$-row-aggregation

$$\rho_\vee(A, k_0) = \quad \begin{array}{c|cccc} & l_1 & l_2 & \cdots & l_n \\ \hline k_0 & \underset{1 \le i \le m}{\vee} (a_{k_i,l_1}) & \underset{1 \le i \le m}{\vee} (a_{k_i,l_2}) & \cdots & \underset{1 \le i \le m}{\vee} (a_{k_i,l_n}) \end{array},$$

$\wedge$-row-aggregation

$$\rho_\wedge(A, k_0) = \quad \begin{array}{c|cccc} & l_1 & l_2 & \cdots & l_n \\ \hline k_0 & \underset{1 \le i \le m}{\wedge} (a_{k_i,l_1}) & \underset{1 \le i \le m}{\wedge} (a_{k_i,l_2}) & \cdots & \underset{1 \le i \le m}{\wedge} (a_{k_i,l_n}) \end{array},$$

$\vee$-column-aggregation

$$\sigma_\vee(A, l_0) = \quad \begin{array}{c|c} & l_0 \\ \hline k_1 & \underset{1 \le j \le n}{\vee} a_{k_1,l_j} \\ \vdots & \vdots \\ k_m & \underset{1 \le j \le n}{\vee} a_{k_m,l_j} \end{array},$$

∧-column-aggregation

$$\sigma_\wedge(A, l_0) = \begin{array}{c|c} & l_0 \\ \hline k_1 & \bigwedge\limits_{1\leq j\leq n} a_{k_1,l_j} \\ \vdots & \vdots \\ k_m & \bigwedge\limits_{1\leq j\leq n} a_{k_m,l_j} \end{array},$$

Now, let us give some examples. Let $\alpha, \ldots, \varphi$ be, e.g., some logical variables. Let the following L-IMs A, B, C, D, E be given

$$A = \begin{array}{c|ccc} & c & d & e \\ \hline a & \alpha & \beta & \gamma \\ b & \delta & \varepsilon & \zeta \\ e & \eta & \theta & \iota \end{array} \qquad B = \begin{array}{c|cc} & c & q \\ \hline a & \kappa & \lambda \\ p & \mu & \nu \\ e & \xi & o \end{array}$$

$$C = \begin{array}{c|ccc} & c & q & e \\ \hline a & \kappa & \lambda & \mu \\ p & \nu & \xi & o \end{array} \quad D = \begin{array}{c|ccc} & c & q & e \\ \hline a & \kappa & \lambda & \mu \\ p & \nu & \xi & o \\ e & \pi & \rho & \sigma \end{array} \quad E = \begin{array}{c|c} & q \\ \hline c & \tau \\ d & \upsilon \\ p & \varphi \end{array}$$

Then, we can construct the following new IMs:

$$A \oplus_\vee B = \begin{array}{c|cccc} & c & d & e & q \\ \hline a & \alpha \vee \kappa & \beta & \gamma & \lambda \\ b & \delta & \varepsilon & \zeta & false \\ e & \eta \vee \xi & \theta & \iota & o \\ p & \mu & false & false & \nu \end{array}$$

$$A \oplus_\wedge C = \begin{array}{c|cccc} & c & d & e & q \\ \hline a & \alpha \wedge \kappa & \beta & \gamma \wedge \mu & \lambda \\ b & \delta & \varepsilon & \zeta & false \\ e & \eta & \theta & \iota & false \\ p & \nu & false & o & \xi \end{array}$$

$$A \oplus_\rightarrow D = \begin{array}{c|cccc} & c & d & e & q \\ \hline a & \alpha \rightarrow \kappa & \beta & \gamma \rightarrow \mu & \lambda \\ b & \delta & \varepsilon & \zeta & false \\ e & \eta \rightarrow \pi & \theta & \iota \rightarrow \sigma & \rho \\ p & \nu & false & o & \xi \end{array}$$

$$A \otimes_\vee B = \begin{array}{c|c} & c \\ \hline a & \alpha \vee \kappa \\ e & \eta \vee \xi \end{array} \quad A \otimes_\wedge C = \begin{array}{c|cc} & c & e \\ \hline a & \alpha \wedge \kappa & \gamma \wedge \mu \end{array}$$

$$A \otimes_\rightarrow D = \begin{array}{c|cc} & c & e \\ \hline a & \alpha \rightarrow \kappa & \gamma \rightarrow \mu \\ e & \eta \rightarrow \pi & \iota \rightarrow \sigma \end{array}$$

$$A \odot_{\vee,\wedge} E = \begin{array}{c|cc} & e & q \\ \hline a & \gamma & (\alpha \wedge \tau) \vee (\beta \wedge \upsilon) \\ b & \zeta & (\delta \wedge \tau) \vee (\varepsilon \wedge \upsilon) \\ p & false & \varphi \\ e & \iota & (\eta \wedge \tau) \vee (\theta \wedge \upsilon) \end{array}$$

More interesting is the following example:

$$\left[*, \frac{e}{f} \right] (A_{(b,*)}) \; \odot_{\wedge,\vee} \; pr_{\{c,d\},\{q,e\}} \left[\frac{a}{c}, \frac{p}{d} \right] D$$

$$= \begin{array}{c|ccc} & c & d & f \\ \hline a & \alpha & \beta & \gamma \\ e & \eta & \theta & \iota \end{array} \odot_{\wedge,\vee} \begin{array}{c|cc} & q & e \\ \hline c & \lambda & \mu \\ d & \xi & 0 \end{array} = \begin{array}{c|ccc} & q & f & e \\ \hline a & (\alpha \vee \lambda) \wedge (\beta \vee \xi) & \gamma & (\alpha \vee \mu) \wedge (\beta \vee 0) \\ e & (\eta \vee \lambda) \wedge (\theta \vee \xi) & \iota & (\eta \vee \mu) \wedge (\theta \vee 0) \end{array}.$$

Finally,

$$A \ominus C = \begin{array}{c|cc} & d & e \\ \hline b & \varepsilon & \zeta \end{array}$$

1.5 Relations Over L-IMs

Let V be an evaluation function that estimates the truth-value of logical variables, propositions or predicates.

The first two relations from Sect. 1.3 keep their form, but the four next relations change their forms, as follows.

The strict relation "inclusion about value" is

$$A \subset_v B \text{ iff } (K = P) \& (L = Q) \& (\forall k \in K)(\forall l \in L)(V(a_{k,l}) < V(b_{k,l})).$$

The non-strict relation "inclusion about value" is

$$A \subseteq_v B \text{ iff } (K = P)\&(L = Q)\&(\forall k \in K)(\forall l \in L)(V(a_{k,l}) \leq V(b_{k,l})).$$

The strict relation "inclusion" is

$$A \subset_* B \text{ iff } (((K \subset P)\&(L \subset Q))\vee((K \subseteq P)\&(L \subset Q))\vee((K \subset P)\&(L \subseteq Q)))$$

$$\&(\forall k \in K)(\forall l \in L)(V(a_{k,l}) < V(b_{k,l})).$$

The non-strict relation "inclusion" is

$$A \subseteq_* B \text{ iff } (K \subseteq P)\&(L \subseteq Q)\&(\forall k \in K)(\forall l \in L)(V(a_{k,l}) \leq V(b_{k,l})).$$

1.6 Operations "Reduction" Over an $\mathcal{R}$-IM, $(0, 1)$-IM and L-IM

Here and below we use symbol "$\perp$" for lack of some component in the separate definitions. In some cases, it is suitable to change this symbol with "0".

Now, we introduce operations $(k, \perp)$- and $(\perp, l)$-reduction of a given IM $A = [K, L, \{a_{k_i,l_j}\}]$:

$$A_{(k,\perp)} = [K - \{k\}, L, \{c_{t_u,v_w}\}]$$

where

$$c_{t_u,v_w} = a_{k_i,l_j} \text{ for } t_u = k_i \in K - \{k\} \text{ and } v_w = l_j \in L$$

and

$$A_{(\perp,l)} = [K, L - \{l\}, \{c_{t_u,v_w}\}],$$

where

$$c_{t_u,v_w} = a_{k_i,l_j} \text{ for } t_u = k_i \in K \text{ and } v_w = l_j \in L - \{l\}.$$

Second, we define

$$A_{(k,l)} = (A_{(k,\perp)})_{(\perp,l)} = (A_{(\perp,l)})_{(k,\perp)},$$

i.e.,

$$A_{(k,l)} = [K - \{k\}, L - \{l\}, \{c_{t_u,v_w}\}],$$

where $c_{t_u,v_w} = a_{k_i,l_j}$ for $t_u = k_i \in K - \{k\}$ and $v_w = l_j \in L - \{l\}$.

Theorem 2 *For every IM A and for every $k_1, k_2 \in K$, $l_1, l_2 \in L$,*

$$(A_{(k_1,l_1)})_{(k_2,l_2)} = (A_{(k_2,l_2)})_{(k_1,l_1)}.$$

Third, let $P = \{k_1, k_2, \ldots, k_s\} \subseteq K$ *and* $Q = \{q_1, q_2, \ldots, q_t\} \subseteq L$. *Then, we define the following three operations:*

$$A_{(P,l)} = (\ldots((A_{(k_1,l)})_{(k_2,l)})\ldots)_{(k_s,l)},$$

$$A_{(k,Q)} = (\ldots((A_{(k,l_1)})_{(k,l_2)})\ldots)_{(k,l_t)},$$

$$A_{(P,Q)} = (\ldots((A_{(p_1,Q)})_{(p_2,Q)})\ldots)_{(p_s,Q)} = (\ldots((A_{(P,q_1)})_{(P,q_2)})\ldots)_{(P,q_t)}.$$

Obviously,

$$A_{(K,L)} = I_{\emptyset} \text{ and }]]A_{(\emptyset,\emptyset)} = A.$$

Theorem 3 *For every two IMs* $A = [K, L, \{a_{k_i,l_j}\}]$, $B = [P, Q, \{b_{p_r,q_s}\}]$:

$$A \subseteq_d B \text{ iff } A = B_{(P-K,Q-L)}.$$

Proof Let $A \subseteq_d B$. Therefore, $K \subseteq P$ and $L \subseteq Q$ and for every $k \in K, l \in L$: $a_{k,l} = b_{k,l}$. From the definition,

$$B_{(P-K,Q-L)} = (\ldots((B_{(p_1,q_1)})_{(p_1,q_2)})\ldots)_{(p_r,q_s)},$$

where $p_1, p_2, \ldots, p_r \in P - K$, i.e., $p_1, p_2, \ldots, p_r \in P$, and $p_1, p_2, \ldots, p_r \notin K$, and $q_1, q_2, \ldots, q_s \in Q - L$, i.e., $q_1, q_2, \ldots, q_s \in Q$, and $q_1, q_2, \ldots, q_s \notin L$. Therefore,

$$B_{(P-K,Q-L)} = [P-(P-K), Q-(Q-L), \{b_{k,l}\}] = [K, L, \{b_{k,l}\}] = [K, L, \{a_{k,l}\}] = A,$$

because by definition the elements of the two IMs, which are indexed by equal symbols, coincide.   □

For the opposite direction we obtain, that if $A = B_{(P-K,Q-L)}$, then

$$A = B_{(P-K,Q-L)} \subseteq_d B_{\emptyset,\emptyset} = B.$$

1.7 Operation "Projection" Over an $\mathcal{R}$-IM, (0, 1)-IM and L-IM

Let $M \subseteq K$ and $N \subseteq L$. Then,

$$pr_{M,N} A = [M, N, \{b_{k_i,l_j}\}],$$

where for each $k_i \in M$ and each $l_j \in N$, $b_{k_i,l_j} = a_{k_i,l_j}$.

Obviously, for every IM A and sets $M_1 \subseteq M_2 \subseteq K$ and $N_1 \subseteq N_2 \subseteq L$ the equality

$$pr_{M_1,N_1} pr_{M_2,N_2} A = pr_{M_1,N_1} A$$

holds.

Theorem 4 *For $M \subseteq K, N \subseteq L$, the equalities*

$$pr_{M,N} A = A_{(K-M,L-N)},$$

$$A_{M,N} = pr_{K-M,L-N} A$$

hold.

1.8 Operation "Substitution" Over an $\mathcal{R}$-IM, (0, 1)-IM and L-IM

Let IM $A = [K, L, \{a_{k,l}\}]$ be given.

First, local substitution over the IM is defined for the couples of indices (p, k) and/or (q, l), respectively, by

$$\left[\frac{p}{k}; \perp\right] A = \left[(K - \{k\}) \cup \{p\}, L, \{a_{k,l}\}\right],$$

$$\left[\perp; \frac{q}{l}\right] A = \left[K, (L - \{l\}) \cup \{q\}, \{a_{k,l}\}\right],$$

Second,

$$\left[\frac{p}{k}; \frac{q}{l}\right] A = \left[\frac{p}{k}; \perp\right]\left[\perp; \frac{q}{l}\right] A,$$

i.e.

$$\left[\frac{p}{k}; \frac{q}{l}\right] A = \left[(K - \{k\}) \cup \{p\}, (L - \{l\}) \cup \{q\}, \{a_{k,l}\}\right].$$

Obviously, for the above indices k, l, p, q:

$$\left[\frac{k}{p}; \perp\right]\left(\left[\perp; \frac{p}{k}\right] A\right) = \left[\frac{l}{q}; \perp\right]\left(\left[\perp; \frac{q}{l}\right] A\right),$$

$$\left[\frac{k}{p}; \frac{l}{q}\right]\left(\left[\frac{p}{k}; \frac{q}{l}\right] A\right) = A.$$

Let the sets of indices $P = \{p_1, p_2, \ldots, p_u\}$, $Q = \{q_1, q_2, \ldots, q_v\}$ be given. Third, for them we define sequentially:

$$\left[\frac{P}{K}; \perp\right] A = \left[\frac{p_1 \ p_2 \ \ldots \ p_u}{k_1 \ k_2 \quad k_u}; \perp\right] A,$$

$$\left[\perp; \frac{Q}{L}\right] A = (\left[\perp; \frac{q_1 \ q_2 \ \ldots \ q_v}{l_1 \ l_2 \quad l_v}\right] A),$$

$$\left[\frac{P}{K}; \frac{Q}{L}\right] A = \left[\frac{P}{K}; \perp\right] (\left[\perp; \frac{Q}{L}\right] A),$$

Obviously, for the sets K, L, P, Q:

$$\left[\frac{K}{P}; \perp\right] (\left[\frac{P}{K}; \perp\right] A) = \left[\perp; \frac{L}{Q}\right] (\left[\perp; \frac{Q}{L}\right] A) = \left[\frac{K}{P}; \frac{L}{Q}\right] (\left[\frac{P}{K}; \frac{Q}{L}\right] A) = A.$$

Theorem 5 *For every four sets of indices P_1, P_2, Q_1, Q_2*

$$\left[\frac{P_2}{P_1}; \frac{Q_2}{Q_1}\right]\left[\frac{P_1}{K}; \frac{Q_1}{L}\right] A = \left[\frac{P_2}{K}; \frac{Q_2}{L}\right] A.$$

1.9 An Example from Number Theory

It is well-known (see e.g., [43]) that each natural number n has a canonical representation $m = \prod_{i=1}^{k} p_i^{\alpha_i}$, where $k, \alpha_1, \alpha_2, \ldots, \alpha_k \geq 1$ are natural numbers and $p_1, p_2, \ldots, p_k$ are different prime numbers. We can always suppose that $p_1 < p_2 < \cdots < p_k$.

Now, we see that m has the following IM-interpretation:

$$IM(m, a) = \begin{array}{c|cccc} & p_1 & p_2 & \cdots & p_k \\ \hline a & \alpha_1 & \alpha_2 & \cdots & \alpha_k \end{array},$$

where "a" is an arbitrary symbol, in a particular case—the same "m".

Obviously, if m is a prime number, its IM-interpretation is

$$IM(p, a) = \begin{array}{c|c} & p \\ \hline a & 1 \end{array}$$

and when $m = pq$ for the prime numbers p and q, its IM-interpretation is

$$IM(m,a) = IM(pq,a) = \frac{\quad}{a}\begin{vmatrix} p & q \\ 1 & 1 \end{vmatrix}.$$

Let us have two natural numbers m and n. In the general case, they have the forms

$$m = \prod_{i=1}^{k+h} p_i^{\alpha_i} \quad\text{and}\quad n = \prod_{i=k+1}^{k+h+g} p_i^{\beta_i}.$$

Therefore,

$$mn = (\prod_{i=1}^{k} p_i^{\alpha_i}).(\prod_{i=k+1}^{k+h} p_i^{\alpha_i+\beta_i}).(\prod_{i=i+h1}^{k+h+g} p_i^{\beta_i}).$$

Now, we see that

$$IM(m,a) \oplus_{(+)} IM(n,a) = \frac{\quad}{a}\begin{vmatrix} p_1 & \cdots & p_k & p_{k+1} & \cdots & p_{k+h} & p_{k+h1} & \cdots & p_{k+h+g} \\ \alpha_1 & \cdots & \alpha_k & \alpha_{k+1}+\beta_{k+1} & \cdots & \alpha_{k+h}+\beta_{k+h} & \beta_{k+h+1} & \cdots & \beta_{k+h+g} \end{vmatrix}$$

$$IM(m.n,a),$$

while if $m = n.s$, then

$$IM(s,a) = IM(\frac{m}{n},a) = IM(m,a) -_{(+)} IM(n,a).$$

On the other hand,

$$IM(m,a) \oplus_{(max)} IM(n,a) = \frac{\quad}{a}\begin{vmatrix} p_1 & \cdots & p_k & p_{k+1} & \cdots & p_{k+h} & p_{k+h+1} & \cdots & p_{k+h+g} \\ \alpha_1 & \cdots & \alpha_k & max(\alpha_{k+1},\beta_{k+1}) & \cdots & max(\alpha_{k+h},\beta_{k+h}) & \beta_{k+h+1} & \cdots & \beta_{k+h+g} \end{vmatrix},$$

that is an IM-interpretation of the least common multiple of m and n. The greatest common divisor of m and n has an IM-interpretation in the form

$$IM(m,a) \otimes_{(min)} IM(n,a) = \frac{\quad}{a}\begin{vmatrix} p_{k+1} & \cdots & p_{k+h} \\ min(\alpha_{k+1},\beta_{k+1}) & \cdots & min(\alpha_{k+h},\beta_{k+h}) \end{vmatrix}.$$

Finally, the IM-interpretation of m^n is

$$IM(m^n,a) = n.IM(m,a) = \frac{\quad}{a}\begin{vmatrix} p_1 & p_2 & \cdots & p_k \\ n\alpha_1 & n\alpha_2 & \cdots & n\alpha_k \end{vmatrix}.$$

The result of operation $IM(m,a) \oplus_{(+)} IM(n,a)$ can be obtained in another way. We can construct the IM

$$A = \begin{array}{c|cccccccc} & p_1 & \cdots & p_k & p_{k+1} & \cdots & p_{k+h} & p_{k+h+1} & \cdots & p_{k+h+g} \\ \hline a_m & \alpha_1 & \cdots & \alpha_k & \alpha_{k+1} & \cdots & \alpha_{k+h} & 0 & \cdots & 0 \\ a_n & 0 & \cdots & 0 & \beta_{k+1} & \cdots & \beta_{k+h} & \beta_{k+h+1} & \cdots & \beta_{k+h+g} \end{array}$$

that represents simultaneously the two numbers m and n and then the IM

$$\rho_{(+)}(A, a) = \begin{array}{c|cccccccc} & p_1 & \cdots & p_k & p_{k+1} & \cdots & p_{k+h} & p_{k+h1} & \cdots & p_{k+h+g} \\ \hline a & \alpha_1 & \cdots & \alpha_k & \alpha_{k+1} + \beta_{k+1} & \cdots & \alpha_{k+h} + \beta_{k+h} & \beta_{k+h+1} & \cdots & \beta_{k+h+g} \end{array}$$

represents $m.n$, the IM

$$\rho_{(max)}(A, a) = \begin{array}{c|cccccccc} & p_1 & \cdots & p_k & p_{k+1} & \cdots & p_{k+h} & p_{k+h+1} & \cdots & p_{k+h+g} \\ \hline a & \alpha_1 & \cdots & \alpha_k & \max(\alpha_{k+1}, \beta_{k+1}) & \cdots & \max(\alpha_{k+h}, \beta_{k+h}) & \beta_{k+h+1} & \cdots & \beta_{k+h+g} \end{array}$$

represents the least common multiple of m and n, while the IM

$$\rho_{(min)}(A, a) = \begin{array}{c|cccccccc} & p_1 & \cdots & p_k & p_{k+1} & \cdots & p_{k+h} & p_{k+h+1} & \cdots & p_{k+h+g} \\ \hline a & 0 & \cdots & 0 & \min(\alpha_{k+1}, \beta_{k+1}) & \cdots & \min(\alpha_{k+h}, \beta_{k+h}) & 0 & \cdots & 0 \end{array}$$

represents the greatest common divisor of m and n.

Now, we can use operation "reduction":

$$\rho_{(min)}(A, a)_{(\perp, \{p_1, \ldots, p_k, p_{k+h+1}, \ldots, p_{k+h+g}\})} = \begin{array}{c|ccc} & p_{k+1} & \cdots & p_{k+h} \\ \hline a & \min(\alpha_{k+1}, \beta_{k+1}) & \cdots & \min(\alpha_{k+h}, \beta_{k+h}) \end{array}$$

or operation "projection":

$$pr_{\{a\}, \{p_{k+h+1}, \ldots, p_{k+h}\}} \rho_{(min)}(A, a) = \begin{array}{c|ccc} & p_{k+1} & \cdots & p_{k+h} \\ \hline a & \min(\alpha_{k+1}, \beta_{k+1}) & \cdots & \min(\alpha_{k+h}, \beta_{k+h}) \end{array}.$$

By the same way, we can represent, e.g., the result of operation $IM(\frac{m}{n}, a) = IM(m, a) -_{(+)} IM(n, a)$.

1.10 An Example from Graph Theory

Let us have the following oriented graph C

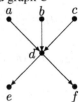

For it, we can construct the $(0, 1)$-IM which is an adjacency matrix of the graph

$$
C = \begin{array}{c|cccccc}
 & a & b & c & d & e & f \\
\hline
a & 0 & 0 & 0 & 1 & 0 & 0 \\
b & 0 & 0 & 0 & 1 & 0 & 0 \\
c & 0 & 0 & 0 & 1 & 0 & 0 \\
d & 0 & 0 & 0 & 0 & 1 & 1 \\
e & 0 & 0 & 0 & 0 & 0 & 0 \\
f & 0 & 0 & 0 & 0 & 0 & 0
\end{array}.
$$

Shortly, we denote this matrix by "Adjacency IM" (AdIM).

Obviously, the columns indexed by a, b, c and the rows, indexed by e, f contain only zeros and do not give any information. So, we can transform the AdIM to the form

$$
D = \begin{array}{c|ccc}
 & d & e & f \\
\hline
a & 1 & 0 & 0 \\
b & 1 & 0 & 0 \\
c & 1 & 0 & 0 \\
d & 0 & 1 & 1
\end{array},
$$

in which the isolated vertices are omitted. This new $(0, 1)$-IM can be called "reduced AdIM". It will be discussed in Chap. 5.

An important question is whether this modification is a correct one. Really, we see that the connections between the immediate neighbouring vertices of the graph are seen, but we must check the basic property of the standard adjacency matrix X, that the elements of the multiplication $X^2 = X \odot_{(\times,+)} X$ represent the connections between the vertices through one step (see e.g., [31]). Using the operation $\odot_{(\max,\min)}$ (see (1.3), p. 3, we obtain for the $(0, 1)$-IM D

$$
D \odot_{(\max,\min)} D = \begin{array}{c|ccc}
 & d & e & f \\
\hline
a & 1 & 0 & 0 \\
b & 1 & 0 & 0 \\
c & 1 & 0 & 0 \\
d & 0 & 1 & 1
\end{array} \odot_{(\max,\min)} \begin{array}{c|ccc}
 & d & e & f \\
\hline
a & 1 & 0 & 0 \\
b & 1 & 0 & 0 \\
c & 1 & 0 & 0 \\
d & 0 & 1 & 1
\end{array} = \begin{array}{c|ccc}
 & d & e & f \\
\hline
a & 0 & 1 & 1 \\
b & 0 & 1 & 1 \\
c & 0 & 1 & 1 \\
d & 0 & 0 & 0
\end{array}
$$

(cf. the first example from Sect. 2.1).

Now, we illustrate the results of the applications of different operations over $(0, 1)$-IMs, that represent some oriented graphs. Let us have the graph E

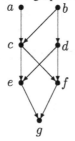

Its $(0, 1)$-IM (in the reduced form, i.e., with omission of the row indexed by g) is

$$E = \begin{array}{c|ccccc} & c & d & e & f & g \\ \hline a & 1 & 0 & 0 & 0 & 0 \\ b & 1 & 1 & 0 & 0 & 0 \\ c & 0 & 0 & 1 & 1 & 0 \\ d & 0 & 0 & 1 & 1 & 0 \\ e & 0 & 0 & 0 & 0 & 1 \\ f & 0 & 0 & 0 & 0 & 1 \end{array} .$$

Then, we calculate:

$$E^2 = E \odot_{\min} E = \begin{array}{c|ccccc} & c & d & e & f & g \\ \hline a & 0 & 0 & 1 & 1 & 0 \\ b & 0 & 0 & 1 & 1 & 0 \\ c & 0 & 0 & 0 & 0 & 1 \\ d & 0 & 0 & 0 & 0 & 1 \\ e & 0 & 0 & 0 & 0 & 0 \\ f & 0 & 0 & 0 & 0 & 0 \end{array} .$$

As above, we can reduce the $(0, 1)$-IM E^2 to the form

$$\begin{array}{c|ccc} & e & f & g \\ \hline a & 1 & 1 & 0 \\ b & 1 & 1 & 0 \\ c & 0 & 0 & 1 \\ d & 0 & 0 & 1 \end{array} .$$

It is interesting to see that

$$E \odot_{\min} E^2 = \begin{array}{c|ccccc} & c & d & e & f & g \\ \hline a & 0 & 0 & 0 & 0 & 1 \\ b & 0 & 0 & 0 & 0 & 1 \\ c & 0 & 0 & 0 & 0 & 0 \\ d & 0 & 0 & 0 & 0 & 0 \\ e & 0 & 0 & 0 & 0 & 0 \\ f & 0 & 0 & 0 & 0 & 0 \end{array} = E^2 \odot_{\min} E.$$

Therefore, $E \odot_{\min} E^2 = E^3 = E^2 \odot_{\min} E$. We see that the resultant $(0, 1)$-IM can be reduced to

$$\begin{array}{c|c} & g \\ \hline a & 1 \\ b & 1 \end{array} .$$

Let us construct a graph, that is a result of operation substitution $\left[\frac{h\ i}{c\ f}; \frac{h\ i}{c\ f}\right]$ over graph C, then we obtain the graph F with IM:

	a	b	h	d	e	i
a	0	0	0	1	0	0
b	0	0	0	1	0	0
h	0	0	0	1	0	0
d	0	0	0	0	1	1
e	0	0	0	0	0	0
i	0	0	0	0	0	0

and with the form:

In the present case, when both index sets coincide, it is suitable to use notation $\left[\frac{h\ i}{c\ f}\right]$.

If we like to unite the graphs E and F, we obtain the following graph with the form

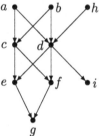

and with (0, 1)-IM

$$E \oplus_{\max} F = \quad$$

	c	d	e	f	g	i
a	1	1	0	0	0	0
b	1	1	0	0	0	0
c	0	0	1	1	0	0
d	0	0	1	1	0	1
e	0	0	0	0	1	0
f	0	0	0	0	1	0
h	0	1	0	0	0	0

The (0, 1)-IM that is a result of operation termwise multiplication with sub-operation "max" over (0, 1)-IMs E and C is

$$E \otimes_{max} F = \begin{array}{c|cc} & d & e \\ \hline a & 1 & 0 \\ b & 1 & 0 \\ d & 0 & 1 \end{array},$$

with sub-operation "min" is

$$E \otimes_{min} F = \begin{array}{c|cc} & d & e \\ \hline a & 0 & 0 \\ b & 1 & 0 \\ d & 0 & 1 \end{array}$$

and it has, respectively, the graph-forms

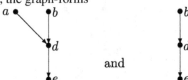

and

In addition, we mention that the graph-representation of the $(0, 1)$-IM

$$E \oplus_{min} F = \begin{array}{c|cccccc} & c & d & e & f & g & i \\ \hline a & 0 & 0 & 0 & 0 & 0 & 0 \\ b & 0 & 1 & 0 & 0 & 0 & 0 \\ c & 0 & 0 & 0 & 0 & 0 & 0 \\ d & 0 & 0 & 1 & 0 & 0 & 0 \\ e & 0 & 0 & 0 & 0 & 0 & 0 \\ f & 0 & 0 & 0 & 0 & 0 & 0 \\ h & 0 & 0 & 0 & 0 & 0 & 0 \end{array}$$

and in the reduced form

$$E \ominus_{min} F = \begin{array}{c|cc} & d & e \\ \hline b & 1 & 0 \\ d & 0 & 1 \end{array}$$

i.e., the graph form is

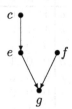

If we like to omit some vertex of a graph, we can do this, using operation "reduction". For example,

$$E_{(d,d)} = \begin{array}{c|cccc} & c & e & f & g \\ \hline a & 1 & 0 & 0 & 0 \\ b & 1 & 0 & 0 & 0 \\ c & 0 & 1 & 1 & 0 \\ e & 0 & 0 & 0 & 1 \\ f & 0 & 0 & 0 & 1 \end{array}.$$

This (0, 1)-IM has a graph-representation

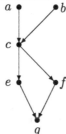

Now, we illustrate operation $\odot_{(\circ,*)}$. Let us have graphs G and H

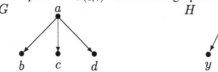

Let us like to add to each of the lower vertices of G new graphs with the form of H and let the vertices of these new graphs be the triples (b, p, q), (c, r, s) and (d, t, u) that will replace vertices (x, y, z), respectively. Then, we obtain the graph

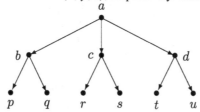

that has the (0, 1)-IM-representation

$$G \odot_{(\circ,*)} \left(\begin{bmatrix} b & p & q \\ \hline x & y & z \end{bmatrix} H \oplus_{(\max)} \begin{bmatrix} c & r & s \\ \hline x & y & z \end{bmatrix} H \oplus_{(\max)} \begin{bmatrix} d & t & q \\ \hline x & y & u \end{bmatrix} H \right).$$

Of course, if, e.g., the graph C is not an oriented, then its AdIM has the form

	a	b	c	d	e	f
a	0	0	0	1	0	0
b	0	0	0	1	0	0
c	0	0	0	1	0	0
d	1	1	1	0	1	1
e	0	0	0	1	0	0
f	0	0	0	1	0	0

while, if we add to it, e.g., the arc (a, e), its AdIM obtains the form

	a	b	c	d	e	f
a	0	0	0	1	1	0
b	0	0	0	1	0	0
c	0	0	0	1	0	0
d	1	1	1	0	1	1
e	1	0	0	1	0	0
f	0	0	0	1	0	0

If the graph has a loop, e.g., (b, b), then its AdIM has the form

	a	b	c	d	e	f
a	0	0	0	1	1	0
b	0	1	0	1	0	0
c	0	0	0	1	0	0
d	1	1	1	0	1	1
e	1	0	0	1	0	0
f	0	0	0	1	0	0

Let us have, for example, the following multi-graph

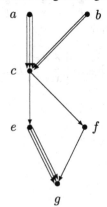

Now, the AdIM has the form

$$
P = \begin{array}{c|cccc}
 & c & e & f & g \\
\hline
a & 3 & 0 & 0 & 0 \\
b & 2 & 0 & 0 & 0 \\
c & 0 & 1 & 1 & 0 \\
e & 0 & 0 & 0 & 3 \\
f & 0 & 0 & 0 & 1
\end{array}.
$$

By similar way we can represent a weighted graph. When we apply to AdIM P operations "Sum-row-aggregation" and "Sum-column-aggregation", we obtain the IMs (they are not $(0, 1)$-IM):

$$
\sigma_{\text{sum}}(P, k) = \begin{array}{c|cccc}
 & c & e & f & g \\
\hline
k & 5 & 1 & 1 & 4
\end{array}
$$

and

$$
\rho_{\text{sum}}(P, l) = \begin{array}{c|c}
 & l \\
\hline
a & 3 \\
b & 2 \\
c & 2 \\
e & 3 \\
f & 1
\end{array}
$$

that shows how many arcs enter and how many arcs leave the individual vertices.

We finish with three **Open problems**

1. Which other operations and relations can be defined over the three types of IMs abd which properties they will have?
2. Which other applications of the IMs can be found in the area of number theory?
3. To represent the basic concepts related to graphs (e.g., path, diameter, etc.) and properties (e.g., planarity, (dis)connectness, symmetry, etc.) of the graphs in IM-form.

Chapter 2
Intuitionistic Fuzzy IMs

Here, following [11, 16], we extend the concept of IM, introducing the concept of an Intuitionistic Fuzzy IM (IFIM) and Extended IFIM (EIFIM).

2.1 Short Remarks on Intuitionistic Fuzziness

Initially, we give some remarks on Intuitionistic Fuzzy Sets (IFSs, see, e.g., [7, 13]) and especially, of their particular case, Intuitionistic Fuzzy Pairs (IFPs; see [26]). The IFP is an object with the form $\langle a, b \rangle$, where $a, b \in [0, 1]$ and $a + b \leq 1$, that is used as an evaluation of some object or process. Its components (a and b) are interpreted as degrees of membership and non-membership, or degrees of validity and non-validity, or degree of correctness and non-correctness, etc.

Let us have two IFPs $x = \langle a, b \rangle$ and $y = \langle c, d \rangle$.

The following relations have been defined in [26]:

$$
\begin{aligned}
x &< y \quad \text{iff} \quad a < c \text{ and } b > d \\
x &\leq y \quad \text{iff} \quad a \leq c \text{ and } b \geq d \\
x &= y \quad \text{iff} \quad a = c \text{ and } b = d \\
x &\geq y \quad \text{iff} \quad a \geq c \text{ and } b \leq d \\
x &> y \quad \text{iff} \quad a > c \text{ and } b < d
\end{aligned}
$$

We define analogous of operations "conjunction" and "disjunction":

$$
\begin{aligned}
x \& y &= \langle \min(a, c), \max(b, d) \rangle \\
x \vee y &= \langle \max(a, c)), \min(b, d) \rangle \\
x + y &= \langle a + c - a.c, b.d \rangle \\
x.y &= \langle a.c, b + d - b.d \rangle \\
x @ y &= \langle \tfrac{a+c}{2}, \tfrac{b+d}{2} \rangle.
\end{aligned}
$$

© Springer International Publishing Switzerland 2014
K.T. Atanassov, *Index Matrices: Towards an Augmented Matrix Calculus*,
Studies in Computational Intelligence 573, DOI 10.1007/978-3-319-10945-9_2

In [13], definitions of 138 operations "implication" and 34 operations "negation" are given. In Table 2.1 the currently existing 45 negations are given. In some of these definitions, we use the functions sg and $\overline{sg}$ that are defined by:

$$\text{sg}(x) = \begin{cases} 1 \text{ if } x > 0 \\ 0 \text{ if } x \leq 0 \end{cases}, \quad \overline{\text{sg}}(x) = \begin{cases} 0 \text{ if } x > 0 \\ 1 \text{ if } x \leq 0 \end{cases}$$

Let a set E be fixed. An Intuitionistic Fuzzy Set (IFS) A in E is an object of the following form (see, e.g., [7, 13]):

$$A = \{\langle x, \mu_A(x), \nu_A(x)\rangle | x \in E\},$$

where functions $\mu_A: E \to [0, 1]$ and $\nu_A: E \to [0, 1]$ define the degree of membership and the degree of non-membership of the element $x \in E$, respectively, and for every $x \in E$:

$$0 \leq \mu_A(x) + \nu_A(x) \leq 1.$$

2.2 IFIMs and EIFIMs

Extending Sect. 1.1, the basic definition of the IFIM is given.

Let I be a fixed set. By IFIM with index sets K and L ($K, L \subset I$), we denote the object:

$$[K, L, \{\langle \mu_{k_i,l_j}, \nu_{k_i,l_j}\rangle\}]$$

$$\equiv \begin{array}{c|ccccc} & l_1 & \cdots & l_j & \cdots & l_n \\ \hline k_1 & \langle \mu_{k_1,l_1}, \nu_{k_1,l_1}\rangle & \cdots & \langle \mu_{k_1,l_j}, \nu_{k_1,l_j}\rangle & \cdots & \langle \mu_{k_1,l_n}, \nu_{k_1,l_n}\rangle \\ \vdots & \vdots & \cdots & \vdots & \cdots & \vdots \\ k_i & \langle \mu_{k_i,l_1}, \nu_{k_i,l_1}\rangle & \cdots & \langle \mu_{k_i,l_j}, \nu_{k_i,l_j}\rangle & \cdots & \langle \mu_{k_i,l_n}, \nu_{k_i,l_n}\rangle \\ \vdots & \vdots & \cdots & \vdots & \cdots & \vdots \\ k_m & \langle \mu_{k_m,l_1}, \nu_{k_m,l_1}\rangle & \cdots & \langle \mu_{k_m,l_j}, \nu_{k_m,l_j}\rangle & \cdots & \langle \mu_{k_m,l_n}, \nu_{k_m,l_n}\rangle \end{array},$$

where for every $1 \leq i \leq m, 1 \leq j \leq n: 0 \leq \mu_{k_i,l_j}, \nu_{k_i,l_j}, \mu_{k_i,l_j} + \nu_{k_i,l_j} \leq 1$.

For brevity, we can mention the above object by $[K, L, \{\langle \mu_{k_i,l_j}, \nu_{k_i,l_j}\rangle\}]$, where

$$K = \{k_1, k_2, \ldots, k_m\},$$

$$L = \{l_1, l_2, \ldots, l_n\},$$

for $1 \leq i \leq m$, and $1 \leq j \leq n$:

$$\mu_{k_i,l_j}, \nu_{k_i,l_j}, \mu_{k_i,l_j} + \nu_{k_i,l_j} \in [0, 1].$$

Table 2.1 The currently existing 45 negations

$\neg_1$	$\langle x, b, a \rangle$
$\neg_2$	$\langle x, \overline{sg}(a), sg(a) \rangle$
$\neg_3$	$\langle x, b, a.b + a^2 \rangle$
$\neg_4$	$\langle x, b, 1 - b \rangle$
$\neg_5$	$\langle x, \overline{sg}(1 - b), sg(1 - b) \rangle$
$\neg_6$	$\langle x, \overline{sg}(1 - b), sg(a) \rangle$
$\neg_7$	$\langle x, \overline{sg}(1 - b), a \rangle$
$\neg_8$	$\langle x, 1 - a, a \rangle$
$\neg_9$	$\langle x, \overline{sg}(a), a \rangle$
$\neg_{10}$	$\langle x, \overline{sg}(1 - b), 1 - b \rangle$
$\neg_{11}$	$\langle x, sg(b), \overline{sg}(b) \rangle$
$\neg_{12}$	$\langle x, b.(b + a), \min(1, a.(b^2 + a + b.a)) \rangle$
$\neg_{13}$	$\langle x, sg(1 - a), \overline{sg}(1 - a) \rangle$
$\neg_{14}$	$\langle x, sg(b), \overline{sg}(1 - a) \rangle$
$\neg_{15}$	$\langle x, \overline{sg}(1 - b), \overline{sg}(1 - a) \rangle$
$\neg_{16}$	$\langle x, \overline{sg}(a), \overline{sg}(1 - a) \rangle$
$\neg_{17}$	$\langle x, \overline{sg}(1 - b), \overline{sg}(b) \rangle$
$\neg_{18}$	$\langle x, b.sg(a), a.sg(b) \rangle$
$\neg_{19}$	$\langle x, b.sg(a), 0 \rangle$
$\neg_{20}$	$\langle x, b, 0 \rangle$
$\neg_{21}$	$\langle x, \min(1 - a, sg(a)), \min(a, sg(1 - a)) \rangle$
$\neg_{22}$	$\langle x, \min(1 - a, sg(a)), 0 \rangle$
$\neg_{23}$	$\langle x, 1 - a, 0 \rangle$
$\neg_{24}$	$\langle x, \min(b, sg(1 - b)), \min(1 - b, sg(b)) \rangle$
$\neg_{25}$	$\langle x, \min(b, sg(1 - b)), 0 \rangle$
$\neg_{26}$	$\langle x, b, a.b + \overline{sg}(1 - a) \rangle$
$\neg_{27}$	$\langle x, 1 - a, a.(1 - a) + \overline{sg}(1 - a) \rangle$
$\neg_{28}$	$\langle x, b, (1 - b).b + \overline{sg}(b) \rangle$
$\neg_{29}$	$\langle x, \max(0, b.a + \overline{sg}(1 - b)), \min(1, a.(b.a + \overline{sg}(1 - b)) + \overline{sg}(1 - a)) \rangle$
$\neg_{30}$	$\langle x, a.b, a.(a.b + \overline{sg}(1 - b)) + \overline{sg}(1 - a) \rangle$
$\neg_{31}$	$\langle x, \max(0, (1 - a).a + \overline{sg}(a)), \min(1, a.((1 - a).a + \overline{sg}(a)) + \overline{sg}(1 - a)) \rangle$
$\neg_{32}$	$\langle x, (1 - a).a, a.((1 - a).a + \overline{sg}(a)) + \overline{sg}(1 - a) \rangle$
$\neg_{33}$	$\langle x, b.(1 - b) + \overline{sg}(1 - b), (1 - b).(b.(1 - b) + \overline{sg}(1 - b)) + \overline{sg}(b)) \rangle$
$\neg_{34}$	$\langle x, b.(1 - b), (1 - b).(b.(1 - b) + \overline{sg}(1 - b)) + \overline{sg}(b) \rangle$
$\neg_{35}$	$\langle \frac{b}{2}, \frac{1+a}{2} \rangle$
$\neg_{36}$	$\langle \frac{b}{3}, \frac{2+a}{3} \rangle$
$\neg_{37}$	$\langle \frac{2b}{3}, \frac{2a+1}{3} \rangle$
$\neg_{38}$	$\langle \frac{1-a}{3}, \frac{2+a}{3} \rangle$
$\neg_{39}$	$\langle \frac{b}{3}, \frac{3-b}{3} \rangle$
$\neg_{40}$	$\langle \frac{2-2a}{3}, \frac{1+2a}{3} \rangle$

(continued)

Table 2.1 (continued)

$\neg 41$	$\left(\frac{2b}{3}, \frac{3-2b}{3}\right)$
$\neg 42, \lambda$	$\left(\frac{b+\lambda-1}{2\lambda}, \frac{a+\lambda}{2\lambda}\right)$, where $\lambda \geq 1$
$\neg 43, \gamma$	$\left(\frac{b+\gamma}{2\gamma+1}, \frac{a+\gamma}{2\gamma+1}\right)$, where $\gamma \geq 1$
$\neg 44, \alpha, \beta$	$\left(\frac{b+\alpha-1}{\alpha+\beta}, \frac{a+\beta}{\alpha+\beta}\right)$, where $\alpha \geq 1$, $\beta \in [0, \alpha]$
$\neg 45, \varepsilon, \eta$	$\langle \min(1, \nu_A(x) + \varepsilon), \max(0, \mu_A(x) - \eta)\rangle$

Now, for above sets K and L, the EIFIM is defined by:

$$[K^*, L^*, \{\langle \mu_{k_i, l_j}, \nu_{k_i, l_j}\rangle\}]$$

$$\equiv
\begin{array}{c|ccccc}
 & l_1, \langle \alpha_1^l, \beta_1^l\rangle & \cdots & l_j, \langle \alpha_j^l, \beta_j^l\rangle & \cdots & l_n, \langle \alpha_n^l, \beta_n^l\rangle \\
\hline
k_1, \langle \alpha_1^k, \beta_1^k\rangle & \langle \mu_{k_1, l_1}, \nu_{k_1, l_1}\rangle & \cdots & \langle \mu_{k_1, l_j}, \nu_{k_1, l_j}\rangle & \cdots & \langle \mu_{k_1, l_n}, \nu_{k_1, l_n}\rangle \\
\vdots & \vdots & \cdots & \vdots & \cdots & \vdots \\
k_i, \langle \alpha_i^k, \beta_i^k\rangle & \langle \mu_{k_i, l_1}, \nu_{k_i, l_1}\rangle & \cdots & \langle \mu_{k_i, l_j}, \nu_{k_i, l_j}\rangle & \cdots & \langle \mu_{k_i, l_n}, \nu_{k_i, l_n}\rangle \\
\vdots & \vdots & \cdots & \vdots & \cdots & \vdots \\
k_m, \langle \alpha_m^k, \beta_m^k\rangle & \langle \mu_{k_m, l_1}, \nu_{k_m, l_1}\rangle & \cdots & \langle \mu_{k_m, l_j}, \nu_{k_m, l_j}\rangle & \cdots & \langle \mu_{k_m, l_n}, \nu_{k_m, l_n}\rangle
\end{array},
$$

where for every $1 \leq i \leq m, 1 \leq j \leq n$:

$$\mu_{k_i, l_j}, \nu_{k_i, l_j}, \mu_{k_i, l_j} + \nu_{k_i, l_j} \in [0, 1],$$

$$\alpha_i^k, \beta_i^k, \alpha_i^k + \beta_i^k \in [0, 1],$$

$$\alpha_j^l, \beta_j^l, \alpha_j^l + \beta_j^l \in [0, 1]$$

and here and below,

$$K^* = \{\langle k_i, \alpha_i^k, \beta_i^k\rangle | k_i \in K\} = \{\langle k_i, \alpha_i^k, \beta_i^k\rangle | 1 \leq i \leq m\},$$

$$L^* = \{\langle l_j, \alpha_j^l, \beta_j^l\rangle | l_j \in L\} = \{\langle l_j, \alpha_j^l, \beta_j^l\rangle | 1 \leq j \leq n\}.$$

Let

$$K^* \subset P^* \text{ iff } (K \subset P) \ \& \ (\forall k_i = p_i \in K)((\alpha_i^k < \alpha_i^p) \ \& \ (\beta_i^k > \beta_i^p)).$$

$$K^* \subseteq P^* \text{ iff } (K \subseteq P) \ \& \ (\forall k_i = p_i \in K)((\alpha_i^k \leq \alpha_i^p) \ \& \ (\beta_i^k \geq \beta_i^p)).$$

All operations and relations over EIFIM must be re-defined, because they have different forms from the above ones. Obviously, the hierarchical operators are not applicable now.

2.3 Standard Operations Over EIFIMs

For the EIFIMs $A = [K^*, L^*, \{\langle \mu_{k_i,l_j}, \nu_{k_i,l_j} \rangle\}]$, $B = [P^*, Q^*, \{\langle \rho_{p_r,q_s}, \sigma_{p_r,q_s} \rangle\}]$, operations that are analogous to the usual matrix operations of addition and multiplication are defined, as well as other specific ones.

Addition-(max,min)

$$A \oplus_{(\max,\min)} B = [T^*, V^*, \{\langle \varphi_{t_u,v_w}, \psi_{t_u,v_w} \rangle\}],$$

where

$$T^* = K^* \cup P^* = \{\langle t_u, \alpha_u^t, \beta_u^t \rangle | t_u \in K \cup P\},$$
$$V^* = L^* \cup Q^* = \{\langle v_w, \alpha_w^v, \beta_w^v \rangle | v_w \in L \cup Q\},$$

$$\alpha_u^t = \begin{cases} \alpha_i^k, & \text{if } t_u \in K - P \\[2mm] \alpha_r^p, & \text{if } t_u \in P - K, \\[2mm] \max(\alpha_i^k, \alpha_r^p), & \text{if } t_u \in K \cap P \end{cases}$$

$$\beta_w^v = \begin{cases} \beta_j^l, & \text{if } v_w \in L - Q \\[2mm] \beta_s^q, & \text{if } t_w \in Q - L, \\[2mm] \min(\beta_j^l, \beta_s^q), & \text{if } t_w \in L \cap Q \end{cases}$$

and

$$\langle \varphi_{t_u,v_w}, \psi_{t_u,v_w} \rangle = \begin{cases} \langle \mu_{k_i,l_j}, \nu_{k_i,l_j} \rangle, & \begin{aligned} &\text{if } t_u = k_i \in K \\ &\text{and } v_w = l_j \in L - Q \\ &\text{or } t_u = k_i \in K - P \\ &\text{and } v_w = l_j \in L; \end{aligned} \\[6mm] \langle \rho_{p_r,q_s}, \sigma_{p_r,q_s} \rangle, & \begin{aligned} &\text{if } t_u = p_r \in P \\ &\text{and } v_w = q_s \in Q - L \\ &\text{or } t_u = p_r \in P - K \\ &\text{and } v_w = q_s \in Q; \end{aligned} \\[6mm] \langle \max(\mu_{k_i,l_j}, \rho_{p_r,q_s}), & \text{if } t_u = k_i = p_r \in K \cap P \\ \min(\nu_{k_i,l_j}, \sigma_{p_r,q_s}) \rangle, & \text{and } v_w = l_j = q_s \in L \cap Q \\[4mm] \langle 0, 1 \rangle, & \text{otherwise} \end{cases}$$

Addition-(min,max)

$$A \oplus_{(\min,\max)} B = [T^*, V^*, \{\langle \varphi_{t_u,v_w}, \psi_{t_u,v_w} \rangle\}],$$

where $T^*, V^*, \alpha_u^t, \beta_w^v$, have the above forms but

$$\langle \varphi_{t_u,v_w}, \psi_{t_u,v_w} \rangle = \begin{cases} \langle \mu_{k_i,l_j}, \nu_{k_i,l_j} \rangle, & \text{if } t_u = k_i \in K \\ & \text{and } v_w = l_j \in L - Q \\ & \text{or } t_u = k_i \in K - P \\ & \text{and } v_w = l_j \in L; \\[2ex] \langle \rho_{p_r,q_s}, \sigma_{p_r,q_s} \rangle, & \text{if } t_u = p_r \in P \\ & \text{and } v_w = q_s \in Q - L \\ & \text{or } t_u = p_r \in P - K \\ & \text{and } v_w = q_s \in Q; \\[2ex] \langle \min(\mu_{k_i,l_j}, \rho_{p_r,q_s}), & \text{if } t_u = k_i = p_r \in K \cap P \\ \max(\nu_{k_i,l_j}, \sigma_{p_r,q_s}) \rangle, & \text{and } v_w = l_j = q_s \in L \cap Q \\[2ex] \langle 0, 1 \rangle, & \text{otherwise} \end{cases}$$

Termwise multiplication-(max,min)

$$A \otimes_{(\max,\min)} B = [T^*, V^*, \{\langle \varphi_{t_u,v_w}, \psi_{t_u,v_w} \rangle\}],$$

where

$$T^* = K^* \cap P^* = \{\langle t_u, \alpha_u^t, \beta_u^t \rangle | t_u \in K \cap P\},$$
$$V^* = L^* \cap Q^* = \{\langle v_w, \alpha_w^v, \beta_w^v \rangle | v_w \in L \cap Q\},$$
$$\alpha_u^t = \min(\alpha_i^k, \alpha_r^p), \text{ for } t_u = k_i = p_r \in K \cap P,$$
$$\beta_w^v = \min(\beta_j^l, \beta_s^q), \text{ for } v_w = l_j = q_s \in L \cap Q$$

and

$$\langle \varphi_{t_u,v_w}, \psi_{t_u,v_w} \rangle = \langle \max(\mu_{k_i,l_j}, \rho_{p_r,q_s}), \min(\nu_{k_i,l_j}, \sigma_{p_r,q_s}) \rangle.$$

Termwise multiplication-(min,max)

$$A \otimes_{(\min,\max)} B = [T^*, V^*, \{\langle \varphi_{t_u,v_w}, \psi_{t_u,v_w} \rangle\}],$$

where $T^*, V^*, \alpha_u^t, \beta_w^v$, have the above forms but

$$\langle \varphi_{t_u,v_w}, \psi_{t_u,v_w} \rangle = \langle \min(\mu_{k_i,l_j}, \rho_{p_r,q_s}), \max(\nu_{k_i,l_j}, \sigma_{p_r,q_s}) \rangle.$$

Multiplication-(max,min)

$$A \odot_{(\max,\min)} B = [T^*, V^*, \langle \varphi_{t_u,v_w}, \psi_{t_u,v_w} \rangle \}],$$

where

$$T^* = (K \cup (P - L))^* = \{ \langle t_u, \alpha_u^t, \beta_u^t \rangle | t_u \in K \cup (P - L) \},$$
$$V^* = (Q \cup (L - P))^* = \{ \langle v_w, \alpha_w^v, \beta_w^v \rangle | v_w \in Q \cup (L - P) \},$$

$$\alpha_u^t = \begin{cases} \alpha_i^k, & \text{if } t_u = k_i \in K \\ \\ \alpha_r^p, & \text{if } t_u = p_r \in P - L \end{cases},$$

$$\beta_w^v = \begin{cases} \beta_j^l, & \text{if } v_w = l_j \in L - P \\ \\ \beta_s^q, & \text{if } t_w = q_s \in Q \end{cases},$$

and

$$\langle \varphi_{t_u,v_w}, \psi_{t_u,v_w} \rangle = \begin{cases} \langle \mu_{k_i,l_j}, \nu_{k_i,l_j} \rangle, & \text{if } t_u = k_i \in K \\ & \text{and } v_w = l_j \in L - P - Q \\ \\ \langle \rho_{p_r,q_s}, \sigma_{p_r,q_s} \rangle, & \text{if } t_u = p_r \in P - L - K \\ & \text{and } v_w = q_s \in Q \\ \\ \langle \max_{l_j=p_r \in L \cap P} (\min(\mu_{k_i,l_j}, \rho_{p_r,q_s})), & \text{if } t_u = k_i \in K \\ & \text{and } v_w = q_s \in Q \\ \\ \min_{l_j=p_r \in L \cap P} (\max(\nu_{k_i,l_j}, \sigma_{p_r,q_s})) \rangle, & \\ \\ \langle 0, 1 \rangle, & \text{otherwise} \end{cases}$$

Multiplication-(min,max)

$$A \odot_{(\min,\max)} B = [T^*, V^*, \langle \varphi_{t_u,v_w}, \psi_{t_u,v_w} \rangle \}],$$

where $T^*, V^*, \alpha_u^t, \beta_w^v$, have the above forms but

$$\langle\varphi_{t_u,v_w}, \psi_{t_u,v_w}\rangle = \begin{cases} \langle\mu_{k_i,l_j}, \nu_{k_i,l_j}\rangle, & \text{if } t_u = k_i \in K \\ & \text{and } v_w = l_j \in L - P - Q \\[2ex] \langle\rho_{p_r,q_s}, \sigma_{p_r,q_s}\rangle, & \text{if } t_u = p_r \in P - L - K \\ & \text{and } v_w = q_s \in Q \\[2ex] \langle\min_{l_j=p_r\in L\cap P}(\max(\mu_{k_i,l_j}, \rho_{p_r,q_s})), & \text{if } t_u = k_i \in K \\ & \text{and } v_w = q_s \in Q \\[2ex] \max_{l_j=p_r\in L\cap P}(\min(\nu_{k_i,l_j}, \sigma_{p_r,q_s}))\rangle, & \\[2ex] \langle 0, 1\rangle, & \text{otherwise} \end{cases}$$

Structural subtraction

$$A \ominus B = [T^*, V^*, \{\langle\varphi_{t_u,v_w}, \psi_{t_u,v_w}\rangle\}],$$

where

$$T^* = (K - P)^* = \{\langle t_u, \alpha_u^t, \beta_u^t\rangle | t_u \in K - P\},$$
$$V^* = (L - Q)^* = \{\langle v_w, \alpha_w^v, \beta_w^v\rangle | v_w \in L - Q\},$$

for the set–theoretic subtraction operation and

$$\alpha_u^t = \alpha_i^k, \text{ for } t_u = k_i \in K - P,$$
$$\beta_w^v = \beta_j^l, \text{ for } v_w = l_j \in L - Q$$

and

$$\langle\varphi_{t_u,v_w}, \psi_{t_u,v_w}\rangle = \langle\mu_{k_i,l_j}, \nu_{k_i,l_j}\rangle, \text{ for } t_u = k_i \in K - P \text{ and } v_w = l_j \in L - Q.$$

Negation of an EIFIM

$$\neg A = [T^*, V^*, \{\neg\langle\mu_{k_i,l_j}, \nu_{k_i,l_j}\rangle\}],$$

where $\neg$ is one of the above intuitionistic fuzzy negations in Table 2.1, or another possible negation.

Termwise subtraction

$$A -_{\max,\min} B = A \oplus_{\max,\min} \neg B,$$

$$A -_{\min,\max} B = A \oplus_{\min,\max} \neg B.$$

Operations "reduction", "projection" and "substitution" coincide with the respective operations defined in Chap. 1 , Sects. 1.6–1.8.

2.4 Relations Over EIFIMs

Let the two EIFIMs $A = [K^*, L^*, \{\langle a_{k,l}, b_{k,l} \rangle\}]$ and $B = [P^*, Q^*, \{\langle c_{p,q}, d_{p,q} \rangle\}]$ be given. We shall introduce the following definitions where $\subset$ and $\subseteq$ denote the relations *"strong inclusion"* and *"weak inclusion"*.

The strict relation "inclusion about dimension" is

$$A \subset_d B \text{ iff } (((K^* \subset P^*) \,\&\, (L^* \subset Q^*)) \vee ((K^* \subseteq P^*) \,\&\, (L^* \subset Q^*))$$

$$\vee((K^* \subset P^*) \,\&\, (L^* \subseteq Q^*))) \,\&\, (\forall k \in K)(\forall l \in L)(\langle a_{k,l}, b_{k,l} \rangle = \langle c_{k,l}, d_{k,l} \rangle).$$

The non-strict relation "inclusion about dimension" is

$$A \subseteq_d B \text{ iff } (K^* \subseteq P^*) \,\&\, (L^* \subseteq Q^*) \,\&\, (\forall k \in K)(\forall l \in L)$$

$$(\langle a_{k,l}, b_{k,l} \rangle = \langle c_{k,l}, d_{k,l} \rangle).$$

The strict relation "inclusion about value" is

$$A \subset_v B \text{ iff } (K^* = P^*) \,\&\, (L^* = Q^*) \,\&\, (\forall k \in K)(\forall l \in L)$$

$$(\langle a_{k,l}, b_{k,l} \rangle < \langle c_{k,l}, d_{k,l} \rangle).$$

The non-strict relation "inclusion about value" is

$$A \subseteq_v B \text{ iff } (K^* = P^*) \,\&\, (L^* = Q^*) \,\&\, (\forall k \in K)(\forall l \in L)$$

$$(\langle a_{k,l}, b_{k,l} \rangle \leq \langle c_{k,l}, d_{k,l} \rangle).$$

The strict relation "inclusion" is

$$A \subset_* B \text{ iff } (((K^* \subset P^*) \,\&\, (L^* \subset Q^*)) \vee ((K^* \subseteq P^*) \,\&\, (L^* \subset Q^*))$$

$$\vee ((K^* \subset P^*) \,\&\, (L^* \subseteq Q^*))) \,\&\, (\forall k \in K)(\forall l \in L)(\langle a_{k,l}, b_{k,l} \rangle < \langle c_{k,l}, d_{k,l} \rangle).$$

The non-strict relation "inclusion" is

$$A \subseteq_* B \text{ iff } (K^* \subseteq P^*) \,\&\, (L^* \subseteq Q^*) \,\&\, (\forall k \in K)(\forall l \in L)$$

$$(\langle a_{k,l}, b_{k,l} \rangle \leq \langle c_{k,l}, d_{k,l} \rangle).$$

2.5 Level Operators Over EIFIMs

Let the EIFIM $A = [K^*, L^*, \{\langle \mu_{k_i, l_j}, \nu_{k_i, l_j} \rangle\}]$ be given.

Let for $i = 1, 2, 3 : \rho_i, \sigma_i, \rho_i + \sigma_i \in [0, 1]$ be fixed numbers.

In [7,13], several level operators are defined. One of them, for a given IFS

$$X = \{\langle x, \mu_X(x), \nu_X(x) \rangle | x \in E\}$$

is

$$N_{\alpha, \beta}(X) = \{\langle x, \mu_X(x), \nu_X(x) \rangle | x \in E \,\&\, \mu_X(x) \geq \alpha \,\&\, \nu_X(x) \leq \beta\},$$

where $\alpha, \beta \in [0, 1]$ are fixed and $\alpha + \beta \leq 1$.

Here, its analogues are introduced. They are three: $N^1_{\rho_1, \sigma_1}$, $N^2_{\rho_2, \sigma_2}$, $N^3_{\rho_3, \sigma_3}$ and affect the K-, L-indices and $\langle \mu_{k_i, l_j}, \nu_{k_i, l_j} \rangle$-elements, respectively. The three operators can be applied over an EIFIM A either sequentially, or simultaneously. In the first case, their forms are

$$N^1_{\rho_1, \sigma_1}(A) = [N_{\rho_1, \sigma_1}(K^*), L^*, \{\langle \varphi_{k_i, l_j}, \psi_{k_i, l_j} \rangle\}],$$

where

$$\langle \varphi_{k_i, l_j}, \psi_{k_i, l_j} \rangle = \langle \mu_{k_i, l_j}, \nu_{k_i, l_j} \rangle$$

only for $\langle k_i, \alpha_i^k, \beta_i^k \rangle \in N_{\rho_1, \sigma_1}(K^*)$ and for each $\langle l_j, \alpha_j^l, \beta_j^l \rangle \in L^*$;

$$N^2_{\rho_2, \sigma_2}(A) = [K^*, N_{\rho_2, \sigma_2}(L^*), \{\langle \varphi_{k_i, l_j}, \psi_{k_i, l_j} \rangle\}],$$

where

$$\langle \varphi_{k_i, l_j}, \psi_{k_i, l_j} \rangle = \langle \mu_{k_i, l_j}, \nu_{k_i, l_j} \rangle$$

for each $\langle k_i, \alpha_i^k, \beta_i^k \rangle \in K^*$ and only for $\langle l_j, \alpha_j^l, \beta_j^l \rangle \in N_{\rho_2, \sigma_2}(L^*)$;

$$N^3_{\rho_3, \sigma_3}(A) = [K^*, L^*, \{\langle \varphi_{k_i, l_j}, \psi_{k_i, l_j} \rangle\}],$$

where

$$\langle \varphi_{k_i, l_j}, \psi_{k_i, l_j} \rangle = \begin{cases} \langle \mu_{k_i, l_j}, \nu_{k_i, l_j} \rangle, & \text{if } \mu_{k_i, l_j} \geq \rho_3 \,\&\, \nu_{k_i, l_j} \leq \sigma_3 \\ \langle 0, 1 \rangle, & \text{otherwise} \end{cases},$$

In the second case, their form is

$$(N^1_{\rho_1, \sigma_1}, N^2_{\rho_2, \sigma_2}, N^3_{\rho_3, \sigma_3})(A) = [N_{\rho_1, \sigma_1}(K^*), N_{\rho_2, \sigma_2}(L^*), \{\langle \varphi_{k_i, l_j}, \psi_{k_i, l_j} \rangle\}],$$

where

$$
\langle \varphi_{k_i,l_j}, \psi_{k_i,l_j} \rangle =
\begin{cases}
\langle \mu_{k_i,l_j}, \nu_{k_i,l_j} \rangle, & \text{if } \langle k_i, \alpha_i^k, \beta_i^k \rangle \in N_{\rho_1,\sigma_1}(K^*) \\
& \text{and } \langle l_j, \alpha_j^l, \beta_j^l \rangle \in N_{\rho_2,\sigma_2}(L^*) \\
& \text{and } \mu_{k_i,l_j} \geq \rho_3 \ \& \ \nu_{k_i,l_j} \leq \sigma_3 \\[2ex]
\langle 0, 1 \rangle, & \text{if } \langle k_i, \alpha_i^k, \beta_i^k \rangle \in N_{\rho_1,\sigma_1}(K^*) \\
& \text{and } \langle l_j, \alpha_j^l, \beta_j^l \rangle \in N_{\rho_2,\sigma_2}(L^*) \\
& \text{and } \mu_{k_i,l_j} < \rho_3 \ \vee \ \nu_{k_i,l_j} > \sigma_3
\end{cases}
,
$$

2.6 Aggregation Operations Over EIFIMs

Let the EIFIM

$$
A =
\begin{array}{c|ccccc}
 & l_1, \langle \alpha_1^l, \beta_1^l \rangle & \dots & l_j, \langle \alpha_j^l, \beta_j^l \rangle & \dots & l_n, \langle \alpha_n^l, \beta_n^l \rangle \\
\hline
k_1, \langle \alpha_1^k, \beta_1^k \rangle & \langle \mu_{k_1,l_1}, \nu_{k_1,l_1} \rangle & \dots & \langle \mu_{k_1,l_j}, \nu_{k_1,l_j} \rangle & \dots & \langle \mu_{k_1,l_n}, \nu_{k_1,l_n} \rangle \\
\vdots & \vdots & \dots & \vdots & \dots & \vdots \\
k_i, \langle \alpha_i^k, \beta_i^k \rangle & \langle \mu_{k_i,l_1}, \nu_{k_i,l_1} \rangle & \dots & \langle \mu_{k_i,l_j}, \nu_{k_i,l_j} \rangle & \dots & \langle \mu_{k_i,l_n}, \nu_{k_i,l_n} \rangle \\
\vdots & \vdots & \dots & \vdots & \dots & \vdots \\
k_m, \langle \alpha_m^k, \beta_m^k \rangle & \langle \mu_{k_m,l_1}, \nu_{k_m,l_1} \rangle & \dots & \langle \mu_{k_m,l_j}, \nu_{k_m,l_j} \rangle & \dots & \langle \mu_{k_m,l_n}, \nu_{k_m,l_n} \rangle
\end{array}
,
$$

be given and let $k_0 \notin K$ and $l_0 \notin L$ be two fixed indices.

Now, we introduce the following 18 operations over it.

(max,max)-row-aggregation

$$
\rho_{(\max,\max)}(A, k_0)
$$

$$
= \frac{}{k_0, \langle \max\limits_{1 \leq i \leq m} \alpha_i^k, \min\limits_{1 \leq i \leq m} \beta_i^k \rangle} \left|
\begin{array}{c}
l_1, \langle \alpha_1^l, \beta_1^l \rangle \qquad \dots \\
\langle \max\limits_{1 \leq i \leq m} \mu_{k_i,l_1}, \min\limits_{1 \leq i \leq m} \nu_{k_i,l_1} \rangle \ \dots
\end{array}
\right.
$$

$$
\begin{array}{c}
\dots \quad l_n, \langle \alpha_n^l, \beta_n^l \rangle \\
\hline
\dots \ \langle \max\limits_{1 \leq i \leq m} \mu_{k_i,l_n}, \min\limits_{1 \leq i \leq m} \nu_{k_i,l_n} \rangle
\end{array},
$$

(max,ave)-row-aggregation

$$
\rho_{(\max,\max)}(A, k_0)
$$

$$
= \frac{}{k_0, \langle \max\limits_{1 \leq i \leq m} \alpha_i^k, \min\limits_{1 \leq i \leq m} \beta_i^k \rangle} \left|
\begin{array}{c}
l_1, \langle \alpha_1^l, \beta_1^l \rangle \qquad \dots \\
\langle \frac{1}{m} \sum\limits_{i=1}^{m} \mu_{k_i,l_1}, \frac{1}{m} \sum\limits_{i=1}^{m} \nu_{k_i,l_1} \rangle \ \dots
\end{array}
\right.
$$

$$\frac{\cdots \qquad l_n, \langle \alpha_n^l, \beta_n^l \rangle}{\cdots \langle \frac{1}{m} \sum_{i=1}^{m} \mu_{k_i, l_n}, \frac{1}{m} \sum_{i=1}^{m} \nu_{k_i, l_n} \rangle},$$

(max,min)-row-aggregation

$$\rho_{(\max, \max)}(A, k_0)$$

$$= \frac{}{k_0, \langle \max_{1 \le i \le m} \alpha_i^k, \min_{1 \le i \le m} \beta_i^k \rangle} \left| \frac{l_1, \langle \alpha_1^l, \beta_1^l \rangle \qquad \cdots}{\langle \min_{1 \le i \le m} \mu_{k_i, l_1}, \max_{1 \le i \le m} \nu_{k_i, l_1} \rangle \cdots} \right.$$

$$\frac{\cdots \qquad l_n, \langle \alpha_n^l, \beta_n^l \rangle}{\cdots \langle \min_{1 \le i \le m} \mu_{k_i, l_n}, \max_{1 \le i \le m} \nu_{k_i, l_n} \rangle},$$

(ave,max)-row-aggregation

$$\rho_{(\min, \max)}(A, k_0)$$

$$= \frac{}{k_0, \langle \frac{1}{m} \sum_{i=1}^{m} \alpha_i^k, \frac{1}{m} \sum_{i=1}^{m} \beta_i^k \rangle} \left| \frac{l_1, \langle \alpha_1^l, \beta_1^l \rangle \qquad \cdots}{\langle \max_{1 \le i \le m} \mu_{k_i, l_1}, \min_{1 \le i \le m} \nu_{k_i, l_1} \rangle \cdots} \right.$$

$$\frac{\cdots \qquad l_n, \langle \alpha_n^l, \beta_n^l \rangle}{\cdots \langle \max_{1 \le i \le m} \mu_{k_i, l_n}, \min_{1 \le i \le m} \nu_{k_i, l_n} \rangle},$$

(ave,ave)-row-aggregation

$$\rho_{(\max, \max)}(A, k_0)$$

$$= \frac{}{k_0, \langle \frac{1}{m} \sum_{i=1}^{m} \alpha_i^k, \frac{1}{m} \sum_{i=1}^{m} \beta_i^k \rangle} \left| \frac{l_1, \langle \alpha_1^l, \beta_1^l \rangle \qquad \cdots}{\langle \frac{1}{m} \sum_{i=1}^{m} \mu_{k_i, l_1}, \frac{1}{m} \sum_{i=1}^{m} \nu_{k_i, l_1} \rangle \cdots} \right.$$

$$\frac{\cdots \qquad l_n, \langle \alpha_n^l, \beta_n^l \rangle}{\cdots \langle \frac{1}{m} \sum_{i=1}^{m} \mu_{k_i, l_n}, \frac{1}{m} \sum_{i=1}^{m} \nu_{k_i, l_n} \rangle},$$

(ave,min)-row-aggregation

$$\rho_{(\max, \max)}(A, k_0)$$

$$= \frac{}{k_0, \langle \frac{1}{m} \sum_{i=1}^{m} \alpha_i^k, \frac{1}{m} \sum_{i=1}^{m} \beta_i^k \rangle} \left| \frac{l_1, \langle \alpha_1^l, \beta_1^l \rangle \qquad \cdots}{\langle \min_{1 \le i \le m} \mu_{k_i, l_1}, \max_{1 \le i \le m} \nu_{k_i, l_1} \rangle \cdots} \right.$$

$$\frac{\cdots \qquad l_n, \langle \alpha_n^l, \beta_n^l \rangle}{\cdots \langle \min\limits_{1 \le i \le m} \mu_{k_i,l_n}, \max\limits_{1 \le i \le m} \nu_{k_i,l_n} \rangle},$$

(min,max)-row-aggregation

$$\rho_{(\min,\max)}(A, k_0)$$

$$= \frac{\qquad \qquad \qquad \Big| \qquad l_1, \langle \alpha_1^l, \beta_1^l \rangle \qquad \cdots}{k_0, \langle \min\limits_{1 \le i \le m} \alpha_i^k, \max\limits_{1 \le i \le m} \beta_i^k \rangle \Big| \langle \max\limits_{1 \le i \le m} \mu_{k_i,l_1}, \min\limits_{1 \le i \le m} \nu_{k_i,l_1} \rangle \cdots}$$

$$\frac{\cdots \qquad l_n, \langle \alpha_n^l, \beta_n^l \rangle}{\cdots \langle \max\limits_{1 \le i \le m} \mu_{k_i,l_n}, \min\limits_{1 \le i \le m} \nu_{k_i,l_n} \rangle},$$

(min,ave)-row-aggregation

$$\rho_{(\max,\max)}(A, k_0)$$

$$= \frac{\qquad \qquad \qquad \Big| \qquad l_1, \langle \alpha_1^l, \beta_1^l \rangle \qquad \cdots}{k_0, \langle \min\limits_{1 \le i \le m} \alpha_i^k, \max\limits_{1 \le i \le m} \beta_i^k \rangle \Big| \langle \frac{1}{m} \sum\limits_{i=1}^{m} \mu_{k_i,l_1}, \frac{1}{m} \sum\limits_{i=1}^{m} \nu_{k_i,l_1} \rangle \cdots}$$

$$\frac{\cdots \qquad l_n, \langle \alpha_n^l, \beta_n^l \rangle}{\cdots \langle \frac{1}{m} \sum\limits_{i=1}^{m} \mu_{k_i,l_n}, \frac{1}{m} \sum\limits_{i=1}^{m} \nu_{k_i,l_n} \rangle},$$

(min,min)-row-aggregation

$$\rho_{(\max,\max)}(A, k_0)$$

$$= \frac{\qquad \qquad \qquad \Big| \qquad l_1, \langle \alpha_1^l, \beta_1^l \rangle \qquad \cdots}{k_0, \langle \min\limits_{1 \le i \le m} \alpha_i^k, \max\limits_{1 \le i \le m} \beta_i^k \rangle \Big| \langle \min\limits_{1 \le i \le m} \mu_{k_i,l_1}, \max\limits_{1 \le i \le m} \nu_{k_i,l_1} \rangle \cdots}$$

$$= \frac{\cdots \qquad l_n, \langle \alpha_n^l, \beta_n^l \rangle}{\cdots \langle \min\limits_{1 \le i \le m} \mu_{k_i,l_n}, \max\limits_{1 \le i \le m} \nu_{k_i,l_n} \rangle},$$

(max,max)-column-aggregation

$$\sigma_{max}(A, l_0) = \frac{\qquad \qquad \Big| \quad l_0, \langle \max\limits_{1 \le i \le m} \alpha_j^l, \min\limits_{1 \le i \le m} \beta_j^l \rangle}{
\begin{array}{c|c}
k_1, \langle \alpha_1^k, \beta_1^k \rangle & \langle \max\limits_{1 \le j \le n} \mu_{k_1,l_j}, \min\limits_{1 \le j \le n} \nu_{k_1,l_j} \rangle \\
\vdots & \vdots \\
k_m, \langle \alpha_m^k, \beta_m^k \rangle & \langle \max\limits_{1 \le j \le n} \mu_{k_m,l_j}, \min\limits_{1 \le j \le n} \nu_{k_m,l_j} \rangle
\end{array}},$$

(max,ave)-column-aggregation

$$\sigma_{max}(A,l_0) = \begin{array}{c|c} & l_0, \langle \max\limits_{1\le i\le m} \alpha^l_j, \min\limits_{1\le i\le m} \beta^l_j \rangle \\ \hline k_1, \langle \alpha^k_1, \beta^k_1 \rangle & \langle \frac{1}{n}\sum\limits_{j=1}^{n}\mu_{k_1,l_j}, \frac{1}{n}\sum\limits_{j=1}^{n}\nu_{k_1,l_j} \rangle \\ \vdots & \vdots \\ k_m, \langle \alpha^k_m, \beta^k_m \rangle & \langle \frac{1}{n}\sum\limits_{j=1}^{n}\mu_{k_m,l_j}, \frac{1}{n}\sum\limits_{j=1}^{n}\nu_{k_m,l_j} \rangle \end{array} \;,$$

(max,min)-column-aggregation

$$\sigma_{max}(A,l_0) = \begin{array}{c|c} & l_0, \langle \max\limits_{1\le i\le m} \alpha^l_j, \min\limits_{1\le i\le m} \beta^l_j \rangle \\ \hline k_1, \langle \alpha^k_1, \beta^k_1 \rangle & \langle \min\limits_{1\le j\le n}\mu_{k_1,l_j}, \max\limits_{1\le j\le n}\nu_{k_1,l_j} \rangle \\ \vdots & \vdots \\ k_m, \langle \alpha^k_m, \beta^k_m \rangle & \langle \min\limits_{1\le j\le n}\mu_{k_m,l_j}, \max\limits_{1\le j\le n}\nu_{k_m,l_j} \rangle \end{array} \;,$$

(ave,max)-column-aggregation

$$\sigma_{max}(A,l_0) = \begin{array}{c|c} & l_0, \langle \frac{1}{n}\sum\limits_{j=1}^{n}\alpha^l_j, \frac{1}{n}\sum\limits_{j=1}^{n}\beta^l_j \rangle \\ \hline k_1, \langle \alpha^k_1, \beta^k_1 \rangle & \langle \max\limits_{1\le j\le n}\mu_{k_1,l_j}, \min\limits_{1\le j\le n}\nu_{k_1,l_j} \rangle \\ \vdots & \vdots \\ k_m, \langle \alpha^k_m, \beta^k_m \rangle & \langle \max\limits_{1\le j\le n}\mu_{k_m,l_j}, \min\limits_{1\le j\le n}\nu_{k_m,l_j} \rangle \end{array} \;,$$

(ave,ave)-column-aggregation

$$\sigma_{max}(A,l_0)$$

$$= \begin{array}{c|c} & l_0, \langle \frac{1}{n}\sum\limits_{j=1}^{n}\alpha^l_j, \frac{1}{n}\sum\limits_{j=1}^{n}\beta^l_j \rangle \\ \hline k_1, \langle \alpha^k_1, \beta^k_1 \rangle & \langle \frac{1}{n}\sum\limits_{j=1}^{n}\mu_{k_1,l_j}, \frac{1}{n}\sum\limits_{j=1}^{n}\nu_{k_1,l_j} \rangle \\ \vdots & \vdots \\ k_m, \langle \alpha^k_m, \beta^k_m \rangle & \langle \frac{1}{n}\sum\limits_{j=1}^{n}\mu_{k_m,l_j}, \frac{1}{n}\sum\limits_{j=1}^{n}\nu_{k_m,l_j} \rangle \end{array} \;.$$

(ave,min)-column-aggregation

$$
\sigma_{max}(A, l_0) =
\begin{array}{c|c}
 & l_0, \langle \frac{1}{n} \sum\limits_{j=1}^{n} \alpha^l_j, \frac{1}{n} \sum\limits_{j=1}^{n} \beta^l_j \rangle \\
\hline
k_1, \langle \alpha^k_1, \beta^k_1 \rangle & \langle \min\limits_{1 \le j \le n} \mu_{k_1,l_j}, \max\limits_{1 \le j \le n} \nu_{k_1,l_j} \rangle \\
\vdots & \vdots \\
k_m, \langle \alpha^k_m, \beta^k_m \rangle & \langle \min\limits_{1 \le j \le n} \mu_{k_m,l_j}, \max\limits_{1 \le j \le n} \nu_{k_m,l_j} \rangle
\end{array}
,
$$

(min,max)-column-aggregation

$$
\sigma_{max}(A, l_0) =
\begin{array}{c|c}
 & l_0, \langle \min\limits_{1 \le i \le m} \alpha^l_j, \max\limits_{1 \le i \le m} \beta^l_j \rangle \\
\hline
k_1, \langle \alpha^k_1, \beta^k_1 \rangle & \langle \max\limits_{1 \le j \le n} \mu_{k_1,l_j}, \min\limits_{1 \le j \le n} \nu_{k_1,l_j} \rangle \\
\vdots & \vdots \\
k_m, \langle \alpha^k_m, \beta^k_m \rangle & \langle \max\limits_{1 \le j \le n} \mu_{k_m,l_j}, \min\limits_{1 \le j \le n} \nu_{k_m,l_j} \rangle
\end{array}
,
$$

(min,ave)-column-aggregation

$$
\sigma_{max}(A, l_0) =
\begin{array}{c|c}
 & l_0, \langle \min\limits_{1 \le i \le m} \alpha^l_j, \max\limits_{1 \le i \le m} \beta^l_j \rangle \\
\hline
k_1, \langle \alpha^k_1, \beta^k_1 \rangle & \langle \frac{1}{n} \sum\limits_{j=1}^{n} \mu_{k_1,l_j}, \frac{1}{n} \sum\limits_{j=1}^{n} \nu_{k_1,l_j} \rangle \\
\vdots & \vdots \\
k_m, \langle \alpha^k_m, \beta^k_m \rangle & \langle \frac{1}{n} \sum\limits_{j=1}^{n} \mu_{k_m,l_j}, \frac{1}{n} \sum\limits_{j=1}^{n} \nu_{k_m,l_j} \rangle
\end{array}
,
$$

(min,min)-column-aggregation

$$
\sigma_{max}(A, l_0) =
\begin{array}{c|c}
 & l_0, \langle \min\limits_{1 \le i \le m} \alpha^l_j, \max\limits_{1 \le i \le m} \beta^l_j \rangle \\
\hline
k_1, \langle \alpha^k_1, \beta^k_1 \rangle & \langle \min\limits_{1 \le j \le n} \mu_{k_1,l_j}, \max\limits_{1 \le j \le n} \nu_{k_1,l_j} \rangle \\
\vdots & \vdots \\
k_m, \langle \alpha^k_m, \beta^k_m \rangle & \langle \min\limits_{1 \le j \le n} \mu_{k_m,l_j}, \max\limits_{1 \le j \le n} \nu_{k_m,l_j} \rangle
\end{array}
.
$$

2.7 Extended Modal Operators Defined Over EIFIMs

Let, as above, $x = \langle a, b \rangle$ be an IFP and let $\alpha, \beta \in [0, 1]$. Some of the extended modal operators defined over x have the following forms (see [13, 26]):

$$F_{\alpha,1-\alpha}(x) = \langle a + \alpha.(1 - a - b), b + \beta.(1 - a - b) \rangle, \quad \text{where } \alpha + \beta \leq 1$$
$$G_{\alpha,\beta}(x) = \langle \alpha.a, \beta.b \rangle$$
$$H_{\alpha,\beta}(x) = \langle \alpha.a, b + \beta.(1 - a - b) \rangle$$
$$H^*_{\alpha,\beta}(x) = \langle \alpha.a, b + \beta.(1 - \alpha.a - b) \rangle$$
$$J_{\alpha,\beta}(x) = \langle a + \alpha.(1 - a - b), \beta.b \rangle$$
$$J^*_{\alpha,\beta}(x) = \langle a + \alpha.(1 - a - \beta.b), \beta.b \rangle$$

and let the level operators have the forms:

$$P_{\alpha,\beta}x = \langle \max(\alpha, a), \min(\beta, b) \rangle$$
$$Q_{\alpha,\beta}x = \langle \min(\alpha, a), \max(\beta, b) \rangle,$$

for $\alpha, \beta \in [0, 1]$ and $\alpha + \beta \leq 1$.

Now we define operators over EIFIMs. Let $O^1_{\alpha_1,\beta_1}$, $O^2_{\alpha_2,\beta_2}$, $O^3_{\alpha_3,\beta_3}$ be three operators and their arguments $\alpha_1, \beta_1, \alpha_2, \beta_2, \alpha_3, \beta_3$ satisfy the respective conditions, given above. The three operators affect the K-, L-indices and $\langle \mu_{k_i,l_j}, \nu_{k_i,l_j} \rangle$-elements, respectively. They can be applied over an EIFIM A sequentially, or simultaneously. In the first case, their forms are

$$(O^1_{\alpha_1,\beta_1}, \perp, \perp)(A)$$

$$= \begin{array}{c|ccc}
 & l_1, \langle \alpha_1^l, \beta_1^l \rangle & \dots & l_n, \langle \alpha_n^l, \beta_n^l \rangle \\
\hline
k_1, O^1_{\alpha_1,\beta_1}(\langle \alpha_1^k, \beta_1^k \rangle) & \langle \mu_{k_1,l_1}, \nu_{k_1,l_1} \rangle & \dots & \langle \mu_{k_1,l_n}, \nu_{k_1,l_n} \rangle \\
\vdots & \vdots & \dots & \vdots \\
k_m, O^1_{\alpha_1,\beta_1}(\langle \alpha_m^k, \beta_m^k \rangle) & \langle \mu_{k_m,l_1}, \nu_{k_m,l_1} \rangle & \dots & \langle \mu_{k_m,l_n}, \nu_{k_m,l_n} \rangle
\end{array},$$

$$(\perp, O^2_{\alpha_2,\beta_2}, \perp)(A)$$

$$= \begin{array}{c|ccc}
 & l_1, O^2_{\alpha_2,\beta_2}(\langle \alpha_1^l, \beta_1^l \rangle) & \dots & l_n, O^2_{\alpha_2,\beta_2}(\langle \alpha_n^l, \beta_n^l \rangle) \\
\hline
k_1, \langle \alpha_1^k, \beta_1^k \rangle & \langle \mu_{k_1,l_1}, \nu_{k_1,l_1} \rangle & \dots & \langle \mu_{k_1,l_n}, \nu_{k_1,l_n} \rangle \\
\vdots & \vdots & \dots & \vdots \\
k_m, \langle \alpha_m^k, \beta_m^k \rangle & \langle \mu_{k_m,l_1}, \nu_{k_m,l_1} \rangle & \dots & \langle \mu_{k_m,l_n}, \nu_{k_m,l_n} \rangle
\end{array},$$

$$(\perp, \perp, O^3_{\alpha_3,\beta_3})(A)$$

$$= \begin{array}{c|ccc} & l_1, \langle \alpha_1^l, \beta_1^l \rangle & \cdots & l_n, \langle \alpha_n^l, \beta_n^l \rangle \\ \hline k_1, \langle \alpha_1^k, \beta_1^k \rangle & O^3_{\alpha_3,\beta_3}(\langle \mu_{k_1,l_1}, \nu_{k_1,l_1} \rangle) & \cdots & O^3_{\alpha_3,\beta_3}(\langle \mu_{k_1,l_n}, \nu_{k_1,l_n} \rangle) \\ \vdots & \vdots & \cdots & \vdots \\ k_m, \langle \alpha_m^k, \beta_m^k \rangle & O^3_{\alpha_3,\beta_3}(\langle \mu_{k_m,l_1}, \nu_{k_m,l_1} \rangle) & \cdots & O^3_{\alpha_3,\beta_3}(\langle \mu_{k_m,l_n}, \nu_{k_m,l_n} \rangle) \end{array}.$$

In the second case, the form of the triple of operators is

$$(O^1_{\alpha_1,\beta_1}, O^2_{\alpha_2,\beta_2}, O^3_{\alpha_3,\beta_3})(A)$$

$$= \begin{array}{c|ccc} & l_1, O^2_{\alpha_2,\beta_2}(\langle \alpha_1^l, \beta_1^l \rangle) & \cdots & l_n, O^2_{\alpha_2,\beta_2}(\langle \alpha_n^l, \beta_n^l \rangle) \\ \hline k_1, O^1_{\alpha_1,\beta_1}(\langle \alpha_1^k, \beta_1^k \rangle) & O^3_{\alpha_3,\beta_3}(\langle \mu_{k_1,l_1}, \nu_{k_1,l_1} \rangle) & \cdots & O^3_{\alpha_3,\beta_3}(\langle \mu_{k_1,l_n}, \nu_{k_1,l_n} \rangle) \\ \vdots & \vdots & \cdots & \vdots \\ k_i, O^1_{\alpha_1,\beta_1}(\langle \alpha_i^k, \beta_i^k \rangle) & O^3_{\alpha_3,\beta_3}(\langle \mu_{k_i,l_1}, \nu_{k_i,l_1} \rangle) & \cdots & O^3_{\alpha_3,\beta_3}(\langle \mu_{k_i,l_n}, \nu_{k_i,l_n} \rangle) \\ \vdots & \vdots & \cdots & \vdots \\ k_m, O^1_{\alpha_1,\beta_1}(\langle \alpha_m^k, \beta_m^k \rangle) & O^3_{\alpha_3,\beta_3}(\langle \mu_{k_m,l_1}, \nu_{k_m,l_1} \rangle) & \cdots & O^3_{\alpha_3,\beta_3}(\langle \mu_{k_m,l_n}, \nu_{k_m,l_n} \rangle) \end{array}.$$

2.8 An Example with Intuitionistic Fuzzy Graphs

Let $V = \{v_1, v_2, \ldots, v_n\}$ be a fixed set of vertices and let each vertex x have a degree of existence $\alpha(x)$ and a degree of non-existence $\beta(x)$. Therefore, we can construct the IFS

$$V^* = \{\langle x, \alpha(x), \beta(x) \rangle | x \in V\} = \{\langle v_i, \alpha(v_i), \beta(v_i) \rangle | 1 \le i \le n\},$$

where for each $x \in V$:

$$\alpha(x), \beta(x), \alpha(x) + \beta(x) \in [0, 1].$$

Let H be a set of arcs between vertices from V. We again can juxtapose to each arc a degree of existence $\mu(x, y)$ and a degree of non-existence $\nu(x, y)$. Therefore, we can construct the new IFS

$$H^* = \{\langle \langle x, y \rangle, \mu(x, y), \nu(x, y) \rangle | x, y \in V\}$$
$$= \{\langle \langle v_i, v_j \rangle, \mu(v_i, v_j), \nu(v_i, v_j) \rangle | 1 \le i, j \le n\},$$

where for each $x, y \in V$:

$$\mu(x, y), \nu(x, y), \mu(x, y) + \nu(x, y) \in [0, 1].$$

Now, for the graph $G = (V, H)$ we can construct the Extended Intuitionistic Fuzzy Graph (EIFG) $G^* = (V^*, H^*)$. It has the following IM-representation:

$$[V^*, V^*, \{\langle \mu(v_i, v_j), \nu(v_i, v_j) \rangle\}]$$

$$= \begin{array}{c|ccc} & v_1, \langle \alpha(v_1), \beta(v_1) \rangle & \dots & v_n, \langle \alpha(v_n), \beta(v_n) \rangle \\ \hline v_1, \langle \alpha(v_1), \beta(v_1) \rangle & \langle \mu_{v_1,v_1}, \nu_{v_1,v_1} \rangle & \dots & \langle \mu_{v_1,v_n}, \nu_{v_1,v_n} \rangle \\ \vdots & \vdots & \dots & \vdots \\ v_i, \langle \alpha(v_i), \beta(v_i) \rangle & \langle \mu_{v_i,v_1}, \nu_{v_i,v_1} \rangle & \dots & \langle \mu_{v_i,v_n}, \nu_{v_i,v_n} \rangle \\ \vdots & \vdots & \dots & \vdots \\ v_n, \langle \alpha(v_n), \beta(v_n) \rangle & \langle \mu_{v_n,v_1}, \nu_{v_n,v_1} \rangle & \dots & \langle \mu_{v_n,v_n}, \nu_{v_n,v_n} \rangle \end{array},$$

where for every $1 \le i \le n, 1 \le j \le n$: $\mu_{v_i,v_j}, \nu_{v_i,v_j} \in [0, 1]$, $\mu_{v_i,v_j} + \nu_{v_i,v_j} \in [0, 1]$, $\alpha(v_i), \beta(v_i) \in [0, 1]$, $\alpha(v_i) + \beta(v_i) \in [0, 1]$.

Let us discuss here for simplicity only the case of oriented graph. Let us denote by $x \to y$ the fact that both vertices x and y are connected by an arc and x is higher than y. Let operation $\circ \in \{+, \max, @, \min, \times\}$.

We call that the EIFG G^* is "well-top-down-(very strong, strong, middle, weak, very weak)-ordered", or shortly, "well-top-down-$\circ$-ordered", if for every two vertices v_i and v_j, such that $v_i \to v_j$, the following inequality holds:

$$\langle \alpha_i, \beta_i \rangle \circ \langle \mu_{v_i,v_j}, \nu_{v_i,v_j} \rangle \ge \langle \alpha_j, \beta_j \rangle.$$

Analogously, we call that the EIFG G^* is "well-bottom-up-(very strong, strong, middle, weak, very weak)-ordered", or shortly, "well-bottom-up-$\circ$-ordered", if for every two vertices v_i and v_j, such that $v_i \to v_j$, the following inequality holds:

$$\langle \alpha_i, \beta_i \rangle \circ \langle \mu_{v_i,v_j}, \nu_{v_i,v_j} \rangle \le \langle \alpha_j, \beta_j \rangle.$$

We illustrate the way for IM-interpretation of the EIFGs by the following example. Let us have the EIFG G^* with the form

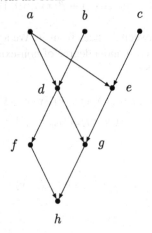

Its H^*-component has the following form (where, obviously, the information about the IFS V^* is included in it):

$$H^* = [\{\langle a, \frac{1}{2}, \frac{1}{3}\rangle, \langle b, \frac{1}{3}, \frac{1}{3}\rangle, \langle c, \frac{1}{3}, \frac{1}{2}\rangle, \langle d, \frac{2}{3}, \frac{1}{8}\rangle, \langle e, \frac{3}{4}, \frac{1}{4}\rangle, \langle f, \frac{1}{10}, \frac{7}{8}\rangle, \langle g, \frac{2}{5}, \frac{3}{5}\rangle,$$

$$\langle h, \frac{1}{5}, \frac{1}{5}\rangle\}, \{\langle a, \frac{1}{2}, \frac{1}{3}\rangle, \langle b, \frac{1}{3}, \frac{1}{3}\rangle, \langle c, \frac{1}{3}, \frac{1}{2}\rangle, \langle d, \frac{2}{3}, \frac{1}{8}\rangle, \langle e, \frac{3}{4}, \frac{1}{4}\rangle, \langle f, \frac{1}{10}, \frac{7}{8}\rangle,$$

$$\langle g, \frac{2}{5}, \frac{3}{5}\rangle, \langle h, \frac{1}{5}, \frac{1}{5}\rangle\}, \{\mu_{x,y}, \nu_{x,y}\}].$$

Now, having in mind the discussion in Sect. 2.2 , we can modify the IM to the form

$$H^* = [\{\langle a, \frac{1}{2}, \frac{1}{3}\rangle, \langle b, \frac{1}{3}, \frac{1}{3}\rangle, \langle c, \frac{1}{3}, \frac{1}{2}\rangle, \langle d, \frac{2}{3}, \frac{1}{8}\rangle, \langle e, \frac{3}{4}, \frac{1}{4}\rangle, \langle f, \frac{1}{10}, \frac{7}{8}\rangle, \langle g, \frac{2}{5}, \frac{3}{5}\rangle\},$$

$$\{\langle d, \frac{2}{3}, \frac{1}{8}\rangle, \langle e, \frac{3}{4}, \frac{1}{4}\rangle, \langle f, \frac{1}{10}, \frac{7}{8}\rangle, \langle g, \frac{2}{5}, \frac{3}{5}\rangle, \langle h, \frac{1}{5}, \frac{1}{5}\rangle\}, \{\mu_{x,y}, \nu_{x,y}\}].$$

The form of the new IM is

	$d, \langle\frac{2}{3}, \frac{1}{8}\rangle$	$e, \langle\frac{3}{4}, \frac{1}{4}\rangle$	$f, \langle\frac{1}{10}, \frac{7}{8}\rangle$	$g, \langle\frac{2}{5}, \frac{3}{5}\rangle$	$h, \langle\frac{1}{5}, \frac{1}{5}\rangle$
$a, \langle\frac{1}{2}, \frac{1}{3}\rangle$	$\langle\frac{3}{4}, \frac{1}{5}\rangle$	$\langle\frac{1}{2}, \frac{1}{4}\rangle$	$\langle 0, 1\rangle$	$\langle 0, 1\rangle$	$\langle 0, 1\rangle$
$b, \langle\frac{1}{3}, \frac{1}{3}\rangle$	$\langle\frac{2}{3}, 0\rangle$	$\langle 0, 1\rangle$	$\langle 0, 1\rangle$	$\langle 0, 1\rangle$	$\langle 0, 1\rangle$
$c, \langle\frac{1}{3}, \frac{1}{2}\rangle$	$\langle 0, 1\rangle$	$\langle\frac{1}{5}, \frac{2}{5}\rangle$	$\langle 0, 1\rangle$	$\langle 0, 1\rangle$	$\langle 0, 1\rangle$
$d, \langle\frac{2}{3}, \frac{1}{8}\rangle$	$\langle 0, 1\rangle$	$\langle 0, 1\rangle$	$\langle\frac{1}{5}, \frac{2}{5}\rangle$	$\langle\frac{3}{4}, \frac{1}{5}\rangle$	$\langle 0, 1\rangle$
$e, \langle\frac{3}{4}, \frac{1}{4}\rangle$	$\langle 0, 1\rangle$	$\langle 0, 1\rangle$	$\langle 0, 1\rangle$	$\langle\frac{2}{3}, \frac{1}{6}\rangle$	$\langle 0, 1\rangle$
$f, \langle\frac{1}{10}, \frac{7}{8}\rangle$	$\langle 0, 1\rangle$	$\langle 0, 1\rangle$	$\langle 0, 1\rangle$	$\langle 0, 1\rangle$	$\langle\frac{1}{3}, \frac{1}{4}\rangle$
$g, \langle\frac{2}{5}, \frac{3}{5}\rangle$	$\langle 0, 1\rangle$	$\langle 0, 1\rangle$	$\langle 0, 1\rangle$	$\langle 0, 1\rangle$	$\langle\frac{3}{5}, \frac{1}{5}\rangle$

Now, we can apply one or more of the level-operators $N^1_{\rho_1,\sigma_1}$, $N^2_{\rho_2,\sigma_2}$, $N^3_{\rho_3,\sigma_3}$ and as a result, the form of the graph will be changed. It is important to mention that in the present case (when the two index sets coincide), the first two level operators must have equal parameters and, therefore, if some vertex has to be omitted from one of both index sets, it will be omitted from the other index set, too. For example, if we apply operator $N^1_{\frac{1}{5},\frac{1}{4}}$ over G^*, we obtain

$$N^1_{\frac13,\frac14}(G^*) =$$

	$d, \langle\frac23,\frac18\rangle$	$e, \langle\frac34,\frac14\rangle$	$g, \langle\frac25,\frac35\rangle$	$h, \langle\frac15,\frac15\rangle$
$a, \langle\frac12,\frac13\rangle$	$\langle\frac34,\frac15\rangle$	$\langle0,1\rangle$	$\langle0,1\rangle$	$\langle0,1\rangle$
$b, \langle\frac13,\frac13\rangle$	$\langle\frac23,0\rangle$	$\langle0,1\rangle$	$\langle0,1\rangle$	$\langle0,1\rangle$
$c, \langle\frac13,\frac12\rangle$	$\langle0,1\rangle$	$\langle\frac15,\frac25\rangle$	$\langle0,1\rangle$	$\langle0,1\rangle$
$d, \langle\frac23,\frac18\rangle$	$\langle0,1\rangle$	$\langle0,1\rangle$	$\langle\frac34,\frac15\rangle$	$\langle0,1\rangle$
$e, \langle\frac34,\frac14\rangle$	$\langle0,1\rangle$	$\langle0,1\rangle$	$\langle\frac25,\frac16\rangle$	$\langle0,1\rangle$
$g, \langle\frac25,\frac35\rangle$	$\langle0,1\rangle$	$\langle0,1\rangle$	$\langle0,1\rangle$	$\langle\frac35,\frac35\rangle$

and the new graph has the form

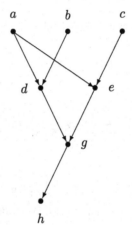

On the other hand, if we can apply, e.g., operator $N^3_{\frac14,\frac13}$ over G^*, we obtain

$$N^3_{\frac14,\frac13}(G^*) =$$

	$d, \langle\frac23,\frac18\rangle$	$e, \langle\frac34,\frac14\rangle$	$f, \langle\frac1{10},\frac78\rangle$	$g, \langle\frac25,\frac35\rangle$	$h, \langle\frac15,\frac15\rangle$
$a, \langle\frac12,\frac13\rangle$	$\langle\frac34,\frac15\rangle$	$\langle\frac12,\frac14\rangle$	$\langle0,1\rangle$	$\langle0,1\rangle$	$\langle0,1\rangle$
$b, \langle\frac13,\frac13\rangle$	$\langle\frac23,0\rangle$	$\langle0,1\rangle$	$\langle0,1\rangle$	$\langle0,1\rangle$	$\langle0,1\rangle$
$c, \langle\frac13,\frac12\rangle$	$\langle0,1\rangle$	$\langle0,1\rangle$	$\langle0,1\rangle$	$\langle0,1\rangle$	$\langle0,1\rangle$
$d, \langle\frac23,\frac18\rangle$	$\langle0,1\rangle$	$\langle0,1\rangle$	$\langle0,1\rangle$	$\langle\frac34,\frac15\rangle$	$\langle0,1\rangle$
$e, \langle\frac34,\frac14\rangle$	$\langle0,1\rangle$	$\langle0,1\rangle$	$\langle0,1\rangle$	$\langle\frac23,\frac16\rangle$	$\langle0,1\rangle$
$f, \langle\frac1{10},\frac78\rangle$	$\langle0,1\rangle$	$\langle0,1\rangle$	$\langle0,1\rangle$	$\langle0,1\rangle$	$\langle\frac13,\frac14\rangle$
$g, \langle\frac25,\frac35\rangle$	$\langle0,1\rangle$	$\langle0,1\rangle$	$\langle0,1\rangle$	$\langle0,1\rangle$	$\langle\frac35,\frac15\rangle$

and the new graph has the form

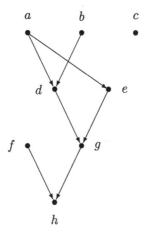

and the IM G^* can be reduced to

	$d, \langle \frac{2}{3}, \frac{1}{8} \rangle$	$e, \langle \frac{3}{4}, \frac{1}{4} \rangle$	$g, \langle \frac{2}{5}, \frac{3}{5} \rangle$	$h, \langle \frac{1}{5}, \frac{1}{5} \rangle$
$a, \langle \frac{1}{2}, \frac{1}{3} \rangle$	$\langle \frac{3}{4}, \frac{1}{5} \rangle$	$\langle \frac{1}{2}, \frac{1}{4} \rangle$	$\langle 0, 1 \rangle$	$\langle 0, 1 \rangle$
$b, \langle \frac{1}{3}, \frac{1}{3} \rangle$	$\langle \frac{2}{3}, 0 \rangle$	$\langle 0, 1 \rangle$	$\langle 0, 1 \rangle$	$\langle 0, 1 \rangle$
$c, \langle \frac{1}{3}, \frac{1}{2} \rangle$	$\langle 0, 1 \rangle$	$\langle 0, 1 \rangle$	$\langle 0, 1 \rangle$	$\langle 0, 1 \rangle$
$d, \langle \frac{2}{3}, \frac{1}{8} \rangle$	$\langle 0, 1 \rangle$	$\langle 0, 1 \rangle$	$\langle \frac{3}{4}, \frac{1}{5} \rangle$	$\langle 0, 1 \rangle$
$e, \langle \frac{3}{4}, \frac{1}{4} \rangle$	$\langle 0, 1 \rangle$	$\langle 0, 1 \rangle$	$\langle \frac{2}{3}, \frac{1}{6} \rangle$	$\langle 0, 1 \rangle$
$f, \langle \frac{1}{10}, \frac{7}{8} \rangle$	$\langle 0, 1 \rangle$	$\langle 0, 1 \rangle$	$\langle 0, 1 \rangle$	$\langle \frac{1}{3}, \frac{1}{4} \rangle$
$g, \langle \frac{2}{5}, \frac{3}{5} \rangle$	$\langle 0, 1 \rangle$	$\langle 0, 1 \rangle$	$\langle 0, 1 \rangle$	$\langle \frac{3}{5}, \frac{1}{5} \rangle$

Obviously, vertex f remains in the first index set, because an arc goes out of it. On the other hand, vertex c here is an isolated one.

The present text was written in the end of 2013 and it was published in [19], when I, as an Editor-in-Chief of the journal "Notes on Intuitionistic Fuzzy Sets", received (in March 2014) the paper of Parvathi Rangasamy [45] that contains very similar ideas. My paper was published in No. 1 for 2014 and Parvathi's paper in No. 2 of the above mentioned journal.

Chapter 3
Extended Index Matrices

The discussed above IMs are, in some sense, extensions of ordinary matrices. Now, we introduce an IM, that includes all of them as particular cases, i.e., it is an extension of the above four types of IMs.

3.1 Definition of an Extended Index Matrix

Let $\mathcal{I}$ be again a fixed set of indices,

$$\mathcal{I}^n = \{\langle i_1, i_2, \ldots, i_n \rangle | (\forall j : 1 \leq j \leq n)(i_j \in \mathcal{I})\}$$

and

$$\mathcal{I}^* = \bigcup_{1 \leq n \leq \infty} \mathcal{I}^n.$$

Let $\mathcal{X}$ be a fixed set of some objects. In the particular cases, they can be either real numbers, or only the numbers 0 or 1, or logical variables, propositions or predicates, etc.

Let operations $\circ, * : \mathcal{X} \times \mathcal{X} \to \mathcal{X}$ be fixed.

An Extended IM (EIM) with index sets K and L ($K, L \subset \mathcal{I}^*$) and elements from set $\mathcal{X}$ is called the object (see, [18]):

$$[K, L, \{a_{k_i,l_j}\}] \equiv
\begin{array}{c|ccccc}
 & l_1 & \cdots & l_j & \cdots & l_n \\
\hline
k_1 & a_{k_1,l_1} & \cdots & a_{k_1,l_j} & \cdots & a_{k_1,l_n} \\
\vdots & \vdots & \cdots & \vdots & \cdots & \vdots \\
k_i & a_{k_i,l_1} & \cdots & a_{k_i,l_j} & \cdots & a_{k_i,l_n} \\
\vdots & \vdots & \cdots & \vdots & \cdots & \vdots \\
k_m & a_{k_m,l_1} & \cdots & a_{k_m,l_j} & \cdots & a_{k_m,l_n}
\end{array},$$

© Springer International Publishing Switzerland 2014

K.T. Atanassov, *Index Matrices: Towards an Augmented Matrix Calculus*,
Studies in Computational Intelligence 573, DOI 10.1007/978-3-319-10945-9_3

where $K = \{k_1, k_2, \ldots, k_m\}$, $L = \{l_1, l_2, \ldots, l_n\}$, for $1 \leq i \leq m$, and $1 \leq j \leq n$: $a_{k_i, l_j} \in \mathcal{X}$.

3.2 Operations Over EIMs

Let in this section, the sets $\mathcal{X}$, $\mathcal{Y}$, $\mathcal{Z}$, $\mathcal{U}$ be fixed. Let operations "$*$" and "$\circ$" be defined so that $* : \mathcal{X} \times \mathcal{Y} \to \mathcal{Z}$ and $\circ : \mathcal{Z} \times \mathcal{Z} \to \mathcal{U}$.

The first six operations from Sect. 1.2 remain valid here without changes.

Now, we see that for operations "addition" and "termwise multiplication",

- in the case of standard, i.e., $\mathcal{R}$-IM, $\mathcal{X} = \mathcal{Y} = \mathcal{R}$, where here and below, $\mathcal{R}$ is the set of the real numbers, operation "$*$" is the standard operation "$+$" or "$\times$" and obviously, $\mathcal{Z} = \mathcal{R}$;
- when $\mathcal{X} = \mathcal{Y} = \{0, 1\}$, operation "$*$" is "max" or "min", and $\mathcal{Z} = \mathcal{X}$;
- when $\mathcal{X} = \mathcal{Y}$ is a set of logical variables, sentences or predicates, then "$*$" is "$\vee$" or "$\wedge$" and $\mathcal{Z} = \mathcal{X}$;
- when

$$\mathcal{X} = \mathcal{Y} = \mathcal{L}^* \equiv \{\langle a, b \rangle | a, b, a + b \in [0, 1]\},$$

then $\mathcal{Z} = \mathcal{X}$ and operation "$*$" is defined for the intuitionistic fuzzy pairs $\langle a, b \rangle$ and $\langle c, d \rangle$, by

$$\langle a, b \rangle * \langle c, d \rangle = \langle \max(a, c), \min(b, d) \rangle$$

or

$$\langle a, b \rangle * \langle c, d \rangle = \langle \min(a, c), \max(b, d) \rangle.$$

In the case of operation "multiplication",

- in the case of standard IM, $\mathcal{X} = \mathcal{Y} = \mathcal{R}$, operation "$*$" is the standard operation "$+$" and operation "$\circ$"—standard operation ".", obviously, $\mathcal{Z} = \mathcal{R}$;
- when $\mathcal{X} = \mathcal{Y} = \{0, 1\}$, operation "$*$" is "max" and "$\circ$"—"min", or opposite, "$*$" is "min" and "$\circ$"—"max", and $\mathcal{Z} = \mathcal{X}$;
- when $\mathcal{X} = \mathcal{Y}$ are a set of logical variables, sentences or predicates, then "$*$" is "$\vee$" and "$\circ$"—"$\wedge$", or vice versa, "$*$" is "$\wedge$" and "$\circ$"—"$\vee$", and $\mathcal{Z} = \mathcal{X}$;
- when $\mathcal{X} = \mathcal{Y} = \mathcal{L}^*$, then $\mathcal{Z} = \mathcal{X}$ and operation $*$ is defined for the intuitionistic fuzzy pairs $\langle a, b \rangle$ and $\langle c, d \rangle$, as above.

In the case of operation "termwise subtraction",

- if $\mathcal{X} = \mathcal{R}$, then the constant $\alpha \in \mathcal{R}$;
- if $\mathcal{X} = \{0, 1\}$, then $\alpha \in \{0, 1\}$;
- when $\mathcal{X} = \mathcal{Y}$ is a set of logical variables, propositions or predicates, then α has sence only when it is an operation "negation".
- when the set $\mathcal{X}$ contains IFPs, then for each one of the above discussed operations over IMs, the operation "$*$" is "max" and "$\circ$" is "min", or vice versa, "$*$" is "min"

and "∘" is "max" and $\mathcal{X} = \mathcal{Y} = \mathcal{Z}$. Only in this case, the form of indices is different: now, it is a triple of index and two numbers in $[0, 1]$ representing the degrees of its validity (existence, etc) and of its non-validity (non-existence, etc).

If $\circ : \mathcal{X} \times \cdots \times \mathcal{X} \rightarrow \mathcal{X}$, the aggregation operations have the forms

∘-**row-aggregation**

$$
\rho_\circ(A, k_0) = \begin{array}{c|cccc}
 & l_1 & l_2 & \cdots & l_n \\
\hline
k_0 & \overset{m}{\underset{i=1}{\circ}} a_{k_i, l_1} & \overset{m}{\underset{i=1}{\circ}} a_{k_i, l_2} & \cdots & \overset{m}{\underset{i=1}{\circ}} a_{k_i, l_n}
\end{array} ,
$$

∘-**column-aggregation**

$$
\sigma_\circ(A, l_0) = \begin{array}{c|c}
 & l_0 \\
\hline
k_1 & \overset{n}{\underset{j=1}{\circ}} a_{k_1, l_j} \\
\vdots & \vdots \\
k_m & \overset{n}{\underset{j=1}{\circ}} a_{k_m, l_j}
\end{array} ,
$$

where, as above, e.g., when ∘ is +, the symbol $\circ_{j=1}^{m}$ will coincide with $\sum_{j=1}^{m}$.

Operations "reduction", "projection" and "substitution" from Sects. 1.6–1.8 are defined over EIMs without changes.

3.3 Relations Over EIMs

Let the two IMs $A = [K, L, \{a_{k,l}\}]$ and $B = [P, Q, \{b_{p,q}\}]$ be given, where $a_{k,l} \in \mathcal{X}, b_{p,q} \in \mathcal{Y}$, and $K, L, P, Q \subset \mathcal{I}$. Let R_s and R_n be a strict and a non-strict relations over $\mathcal{X} \times \mathcal{Y}$, respectively.

We introduce the following definitions, where $\subset$ and $\subseteq$ denote the relations "*strong inclusion*" and "*weak inclusion*" over standard sets.

The first two relations from Sect. 1.3 keep their form, while, the rest four definitions are valid only if there are two relations R_s and R_n—strict and non-strict, defined over $\mathcal{X} \times \mathcal{Y}$.

Then, for EIMs, the definitions for relations over EIMs obtain respectively the forms

The strict relation "inclusion about value" is

$$A \subset_v B \text{ iff } (K = P)\&(L = Q)\&(\forall k \in K)(\forall l \in L)(R_s(a_{k,l}, b_{k,l})).$$

The non-strict relation "inclusion about value" is

$$A \subseteq_v B \text{ iff } (K = P)\&(L = Q)\&(\forall k \in K)(\forall l \in L)(R_n(a_{k,l}, b_{k,l})).$$

The strict relation "inclusion" is

$$A \subset_* B \text{ iff } (((K \subset P) \& (L \subset Q)) \vee ((K \subseteq P) \& (L \subset Q))$$

$$\vee ((K \subset P) \& (L \subseteq Q))) \& (\forall k \in K)(\forall l \in L)(R_s(a_{k,l}, b_{k,l})).$$

The non-strict relation "inclusion" is

$$A \subseteq_* B \text{ iff } (K \subseteq P)\&(L \subseteq Q)\&(\forall k \in K)(\forall l \in L)(R_n(a_{k,l}, b_{k,l})).$$

3.4 Hierarchical Operators Over EIMs

In [10, 14], two hierarchical operators are defined. They are applicable on EIM, when their elements are not only numbers, variables, etc, but when they also can be whole (new) IMs.

Let A be an ordinary IM and let its element a_{k_f,e_g} be an IM by itself:

$$a_{k_f,l_g} = [P, Q, \{b_{p_r,q_s}\}],$$

where

$$K \cap P = L \cap Q = \emptyset.$$

Here, we will introduce the first hierarchical operator:

$$A|(a_{k_f,l_g}) = [(K - \{k_f\}) \cup P, (L - \{l_g\}) \cup Q, \{c_{t_u,v_w}\}],$$

where

$$c_{t_u,v_w} = \begin{cases} a_{k_i,l_j}, & \text{if } t_u = k_i \in K - \{k_f\} \text{ and } v_w = l_j \in L - \{l_g\} \\ b_{p_r,q_s}, & \text{if } t_u = p_r \in P \text{ and } v_w = q_s \in Q \\ 0, & \text{otherwise} \end{cases}.$$

Let us assume that in the case when a_{k_f,l_g} is not an element of IM A, then

$$A|(a_{k_f,l_g}) = A.$$

Let for $i = 1, 2, \ldots, m$:

$$a^i_{k_{i,f},l_{i,g}} = [P_i, Q_i, \{b^i_{p_{i,r},q_{i,s}}\}],$$

where for every i, j $(1 \le i < j \le m)$:

$$P_i \cap P_j = Q_i \cap Q_j = \emptyset,$$
$$P_i \cap K = Q_i \cap L = \emptyset.$$

Then, for $k_{1,f}, k_{2,f}, \ldots, k_{m,f} \in K$ and $l_{1,g}, l_{2,g}, \ldots, l_{m,g} \in L$:

$$A|(a^1_{k_{1,f},l_{1,g}}, a^2_{k_{2,f},l_{2,g}}, \ldots, a^m_{k_{m,f},l_{m,g}})$$
$$= (\ldots((A|(a^1_{k_{1,f},l_{1,g}}))|(a^2_{k_{2,f},l_{2,g}}))\ldots)|(a^m_{k_{m,f},l_{m,g}}).$$

Theorem 6 *Let the IM A be given and let for $i = 1, 2$: $k_{1,f} \ne k_{2,f}$ and $l_{1,g} \ne l_{2,g}$ and*

$$a^i_{k_{i,f},l_{i,g}} = [P_i, Q_i, \{b^i_{p_{i,r},q_{i,s}}\}],$$

where

$$P_1 \cap P_2 = Q_1 \cap Q_2 = \emptyset,$$
$$P_i \cap K = Q_i \cap L = \emptyset.$$

Then,

$$A|(a^1_{k_{1,f},l_{1,g}}, a^2_{k_{2,f},l_{2,g}}) = A|(a^2_{k_{2,f},l_{2,g}}, a^1_{k_{1,f},l_{1,g}}).$$

As it is mentioned in [14], the condition $k_{1,f} \ne k_{2,f}$ and $l_{1,g} \ne l_{2,g}$ is important (it was omitted in [16]), because if we have two elements of a given IM, that are IMs and that belong to one row or column, this will generate problems.

Let us give an example. Let the IM A have the form

$$A = \begin{array}{c|ccc} & l_1 & l_2 & l_3 \\ \hline k_1 & a_{1,1} & a_{1,2} & a_{1,3} \\ k_2 & a_{2,1} & a_{2,2} & a_{2,3} \end{array}.$$

Let elements $a_{2,1}$ and $a_{2,2}$ be IMs and let

$$a_{2,1} = \begin{array}{c|c} & n_1 \\ \hline m_1 & b_{1,1} \\ m_2 & b_{2,1} \end{array}, \qquad a_{2,2} = \begin{array}{c|cc} & q_1 & q_2 \\ \hline p_1 & c_{1,1} & c_{1,2} \end{array}.$$

Then,

$$A|(a_{2,1}, a_{2,2}) = B|(a_{2,2}) = B,$$

because $a_{2,1}$ is not an element of B, where

$$B = \begin{array}{c|ccc} & n_1 & l_2 & l_3 \\ \hline k_1 & 0 & a_{1,2} & a_{1,3} \\ m_1 & b_{2,1} & 0 & 0 \\ m_2 & b_{2,2} & 0 & 0 \end{array}$$

and

$$A|(a_{2,2}, a_{2,1}) = C|(a_{2,1}) = C,$$

because $a_{2,2}$ is not an element of C, where

$$C = \begin{array}{c|cccc} & l_1 & q_1 & q_2 & l_3 \\ \hline k_1 & a_{1,1} & 0 & 0 & a_{1,3} \\ p_1 & 0 & c_{1,1} & c_{1,2} & 0 \end{array},$$

where, obviously, $B \neq C$.

Let A and a_{k_f, l_g} be as above, let b_{m_d, n_e} be the element of the IM a_{k_f, l_g}, and let

$$b_{m_d, n_e} = [R, S, \{c_{t_u, v_w}\}],$$

where

$$K \cap R = L \cap S = P \cap R = Q \cap S = K \cap P = L \cap Q = \emptyset.$$

Then,

$$(A|(a_{k_f, l_g}))|(b_{m_d, n_e})$$
$$= [(K - \{k_f\}) \cup (P - \{m_d\}) \cup R, (L - \{l_g\}) \cup (Q - \{n_e\} \cup S \{\alpha_{\beta_\gamma, \delta_\varepsilon}\}],$$

where

$$
\alpha_{\beta_\gamma,\delta_\varepsilon} = \begin{cases} a_{k_i,l_j}, & \text{if } \beta_\gamma = k_i \in K - \{k_f\} \text{ and } \delta_\varepsilon = l_j \in L - \{l_g\} \\[2mm] b_{p_r,q_s}, & \text{if } \beta_\gamma = p_r \in P - \{m_d\} \text{ and } \delta_\varepsilon = q_s \in Q - \{n_e\} \\[2mm] c_{t_u,v_w}, & \text{if } \beta_\gamma = t_u \in R \text{ and } \delta_\varepsilon = v_w \in S \\[2mm] 0, & \text{otherwise} \end{cases}
$$

Theorem 7 *For the above EIMs A, a_{k_f,l_g} and b_{m_d,n_e}*

$$
(A|(a_{k_f,l_g}))|(b_{m_d,n_e}) = A|((a_{k_f,l_g})|(b_{m_d,n_e})).
$$

From the first definition of a hierarchical operator it follows that

$$
A|(a_{k_f,l_g})
$$

$$
=
$$

	l_1	$\cdots$	l_{g-1}	q_1	$\cdots$	q_u	l_{g+1}	$\cdots$	l_n
k_1	a_{k_1,l_1}	$\cdots$	$a_{k_1,l_{g-1}}$	0	$\cdots$	0	$a_{k_1,l_{g+1}}$	$\cdots$	a_{k_1,l_n}
$\vdots$	$\vdots$	$\vdots$	$\vdots$	$\vdots$	$\vdots$	$\vdots$	$\vdots$	$\vdots$	$\vdots$
k_{f-1}	a_{k_{f-1},l_1}	$\cdots$	$a_{k_{f-1},l_{g-1}}$	0	$\cdots$	0	$a_{k_{f-1},l_{g+1}}$	$\cdots$	a_{k_{f-1},l_n}
p_1	0	$\cdots$	0	b_{p_1,q_1}	$\cdots$	b_{p_1,q_v}	0	$\cdots$	0
$\vdots$	$\vdots$	$\vdots$	$\vdots$	$\vdots$	$\vdots$	$\vdots$	$\vdots$	$\vdots$	$\vdots$
p_u	0	$\cdots$	0	b_{p_u,q_1}	$\cdots$	b_{p_u,q_v}	0	$\cdots$	0
k_{f+1}	a_{k_{f+1},l_1}	$\cdots$	$a_{k_{f+1},l_{g-1}}$	0	$\cdots$	0	$a_{k_{f+1},l_{g+1}}$	$\cdots$	a_{k_{f+1},l_n}
$\vdots$	$\vdots$	$\vdots$	$\vdots$	$\vdots$	$\vdots$	$\vdots$	$\vdots$	$\vdots$	$\vdots$
k_m	a_{k_m,l_1}	$\cdots$	$a_{k_m,l_{g-1}}$	0	$\cdots$	0	$a_{k_m,l_{g+1}}$	$\cdots$	a_{k_m,l_n}

From this form of the IM $A|(a_{k_f,l_g})$ we see that for the hierarchical operator the following equality holds.

Theorem 8 *Let $A = [K, L, \{a_{k_i,l_j}\}]$ be an IM and let $a_{k_f,l_g} = [P, Q, \{b_{p_r,q_s}\}]$ be its element. Then*

$$
A|(a_{k_f,l_g}) = (A \ominus [\{k_f\}, \{l_g\}, \{0\}]) \oplus a_{k_f,l_g}.
$$

We see that the elements $a_{k_f,l_1}, a_{k_f,l_2}, \ldots, a_{k_f,l_{g-1}}, a_{k_f,l_{g+1}}, \ldots, a_{k_f,l_n}$ in the IM A now are replaced by "0". Therefore, as a result of this operator, information is being lost.

Below, we modify the first hierarchical operator, so that all the information from the IMs, participating in it, be kept. The new—second—form of this operator for the above defined IM A and its fixed element a_{k_f,l_g}, is

$$
A|^*(a_{k_f,l_g})
$$

$$= \begin{array}{c|ccccccccc}
 & l_1 & \cdots & l_{g-1} & q_1 & \cdots & q_u & l_{g+1} & \cdots & l_n \\
\hline
k_1 & a_{k_1,l_1} & \cdots & a_{k_1,l_{g-1}} & a_{k_1,l_g} & \cdots & a_{k_1,l_g} & a_{k_1,l_{g+1}} & \cdots & a_{k_1,l_n} \\
\vdots & \vdots & \vdots & \vdots & \vdots & \vdots & \vdots & \vdots & \vdots & \vdots \\
k_{f-1} & a_{k_{f-1},l_1} & \cdots & a_{k_{f-1},l_{g-1}} & a_{k_{f-1},l_g} & \cdots & a_{k_{f-1},l_g} & a_{k_{f-1},l_{g+1}} & \cdots & a_{k_{f-1},l_n} \\
p_1 & a_{k_f,l_1} & \cdots & a_{k_f,l_{g-1}} & b_{p_1,q_1} & \cdots & b_{p_1,q_v} & a_{k_f,l_{g+1}} & \cdots & a_{k_f,l_n} \\
\vdots & \vdots & \vdots & \vdots & \vdots & \vdots & \vdots & \vdots & \vdots & \vdots \\
p_u & a_{k_f,l_1} & \cdots & a_{k_f,l_{g-1}} & b_{p_u,q_1} & \cdots & b_{p_u,q_v} & a_{k_f,l_{g+1}} & \cdots & a_{k_f,l_n} \\
k_{f+1} & a_{k_{f+1},l_1} & \cdots & a_{k_{f+1},l_{g-1}} & a_{k_{f+1},l_g} & \cdots & a_{k_{f+1},l_g} & a_{k_{f+1},l_{g+1}} & \cdots & a_{k_{f+1},l_n} \\
\vdots & \vdots & \vdots & \vdots & \vdots & \vdots & \vdots & \vdots & \vdots & \vdots \\
k_m & a_{k_m,l_1} & \cdots & a_{k_m,l_{g-1}} & a_{k_m,l_g} & \cdots & a_{k_m,l_g} & a_{k_m,l_{g+1}} & \cdots & a_{k_m,l_n}
\end{array}.$$

Now, the following assertion is valid.

Theorem 9 Let $A = [K, L, \{a_{k_i,l_j}\}]$ be an IM and let $a_{k_f,l_g} = [P, Q, \{b_{p_r,q_s}\}]$ be its element. Then

$$A|^*(a_{k_f,l_g})$$
$$= (A \ominus [\{k_f\}, \{l_g\}, \{0\}]) \oplus a_{k_f,l_g} \oplus [P, L - \{l_g\}, \{c_{x,l_j}\}] \oplus [K - \{k_f\}, Q, \{d_{k_i,y}\}],$$

where for each $t \in P$ and for each $l_j \in L - \{l_g\}$,

$$c_{x,l_j} = a_{k_f,l_j}$$

and for each $k_i \in K - \{k_f\}$ and for each $y \in Q$,

$$d_{k_i,y} = a_{k_i,l_g}.$$

We can give other representations of IMs $A|(a_{k_f,l_g})$ and $A|^*(a_{k_f,l_g})$, using other operations defined over IMs.

The following equalities are valid.

Theorem 10 Let $A = [K, L, \{a_{k_i,l_j}\}]$ be an IM and let $a_{k_f,l_g} = [P, Q, \{b_{p_r,q_s}\}]$ be its element. Then

$$A|(a_{k_f,l_g}) = pr_{K-\{k_f\},L-\{l_g\}}A \oplus a_{k_f,l_g},$$
$$A|^*(a_{k_f,l_g})$$
$$= pr_{K-\{k_f\},L-\{l_g\}}A \oplus a_{k_f,l_g} \oplus [P, L - \{l_g\}, \{c_{x,l_j}\}] \oplus [K - \{k_f\}, Q, \{d_{k_i,y}\}],$$

where for each $x \in P$ and for each $l_j \in L - \{l_g\}$,

$$c_{x,l_j} = a_{k_f,l_j}$$

and for each $k_i \in K - \{k_f\}$ and for each $y \in Q$,

$$d_{k_i,y} = a_{k_i,l_g}.$$

Theorem 11 *Let $A = [K, L, \{a_{k_i,l_j}\}]$ be an IM and let $a_{k_f,l_g} = [P, Q, \{b_{p_r,q_s}\}]$ be its element. Then*

$$A|(a_{k_f,l_g}) = A_{(k_f,l_g)} \oplus a_{k_f,l_g},$$
$$A|^*(a_{k_f,l_g}) = A_{(k_f,l_g)} \oplus a_{k_f,l_g} \oplus [P, L - \{l_g\}, \{c_{x,l_j}\}] \oplus [K - \{k_f\}, Q, \{d_{k_i,y}\}],$$

where for each $x \in P$ and for each $l_j \in L - \{l_g\}$,

$$c_{x,l_j} = a_{k_f,l_j}$$

and for each $k_i \in K - \{k_f\}$ and for each $y \in Q$,

$$d_{k_i,y} = a_{k_i,l_g}.$$

Now, we can see that the newly introduced types of IMs, namely, IFIMs, EIFIMs, TIFIMs and ETIFIMs can be represented as EIMs, too. Indeed, if we put

$$\overline{\mathcal{I}} = \mathcal{I} \times [0, 1] \times [0, 1]$$

and

$$\overline{\mathcal{X}} = \mathcal{X} \times [0, 1] \times [0, 1],$$

then we directly see that the IFIMs and EIFIMs can be represented as EIMs, while, for the sets

$$\overline{\overline{\mathcal{I}}} = \mathcal{I} \times [0, 1] \times [0, 1] \times \mathcal{T}$$

and

$$\overline{\overline{\mathcal{X}}} = \mathcal{X} \times [0, 1] \times [0, 1] \times \mathcal{T},$$

the TIFIMs and ETIFIMs can be represented as sets of EIMs.

3.5 New Operations Over EIMs

Now, we introduce some new (non-standard) operations over EIMs.

Let index set $\mathcal{I}$ and set $\mathcal{X}$ be fixed and let the EIMs $A_1, A_2, \ldots, A_n$ over both sets be given.

Let for s ($1 \leq s \leq n$):

$$(\forall p, q)(1 \leq p < q \leq n)(K^p \cap K^q = L^p \cap L^q = \emptyset)$$

and

$$A_s = [K^s, L^s, \{a^s_{k_i, l_j}\}] = \begin{array}{c|cccccc} & l_{s,1} & \cdots & l_{s,j} & \cdots & l_{s,n_s} \\ \hline k_{s,1} & a_{k_{s,1},l_{s,1}} & \cdots & a_{k_{s,1},l_{s,j}} & \cdots & a_{k_{s,1},l_{s,n_s}} \\ \vdots & \vdots & \cdots & \vdots & \cdots & \vdots \\ k_{s,i} & a_{k_{s,i},l_{s,1}} & \cdots & a_{k_{s,i},l_{s,j}} & \cdots & a_{k_{s,i},l_{s,n_s}} \\ \vdots & \vdots & \cdots & \vdots & \cdots & \vdots \\ k_{s,m} & a_{k_{s,m},l_{s,1}} & \cdots & a_{k_{s,m},l_{s,j}} & \cdots & a_{k_{s,m},l_{s,n_s}} \end{array}.$$

The first new operation that we call **"composition"** is defined by

$$\flat\{A_s | 1 \leq s \leq n\} = [\bigcup_{s=1}^n K^s, \bigcup_{s=1}^n L^s, \{\langle c_{1,t_{1,u},v_{1,w}}, c_{2,t_{2,u},v_{2,w}}, \ldots, c_{n,t_{n,u},v_{n,w}} \rangle\}],$$

where for r ($1 \leq r \leq n$):

$$c_{r,t_u,v_w} = \begin{cases} a_{r,k_i,l_j}, & \text{if } t_u = k_i \in K^r \text{ and } v_w = l_j \in L^r \\ \perp, & \text{otherwise} \end{cases}$$

Therefore, it is composed of a new EIM on the basis of n EIMs. The new EIM contains n-dimensional vectors as elements. By this reason, we define function dim, giving the dimensionality of the elements of the EIM A, i.e., for the above EIM, the equality $dim(A) = n$ holds.

The second new operator, that we call **"automatic reduction"** is defined for a given EIM A by

$$@(A) = [P, Q, \{b_{p_r,q_s}\}],$$

where $P \subseteq K$, $Q \subseteq L$ are index sets with the following property:

$$(\forall k \in K - P)(\forall l \in L)(a_{k_i,l_j} = \perp) \ \& \ (\forall k \in K)(\forall l \in L - Q)(a_{k_i,l_j} = \perp)$$

$$\& (\forall p_r = a_i \in P)(\forall q_s = b_j \in Q)(b_{p_r,q_s} = a_{k_i,l_j}).$$

For example, if

$$A = \begin{array}{c|cccc} & d & e & f & g \\ \hline a & 1 & 2 & \perp & 3 \\ b & \perp & \perp & \perp & \perp \\ c & 4 & 5 & \perp & \perp \end{array},$$

then

$$@A = \begin{array}{c|ccc} & d & e & g \\ \hline a & 1 & 2 & 3 \\ c & 4 & 5 & \perp \end{array}.$$

Let $\mathcal{X}$ be a set of n-dimensional vectors and A be an EIM with elements from set $\mathcal{X}$. Then

$$Pr_s(A) = \begin{cases} I_\emptyset, & \text{if } s \leq 0 \text{ or } s > n \\ \\ A_s, & \text{otherwise} \end{cases},$$

where

$$A_s = [K^s, L^s, \{a^s_{k_i,l_j}\}]$$

and $a^s_{k_i,l_j}$ is the s-th component of vector $\langle a^1_{k_i,l_j}, a^2_{k_i,l_j}, \ldots, a^n_{k_i,l_j} \rangle$ that is an element of A.

Now, we can define an operation, that is in some sense opposite of operation $\flat$. It has the form

$$\sharp(A) = \{@Pr_s(A) | 1 \leq s \leq n\}.$$

We give an example. Let the EIMs A_1, A_2 have the forms

$$A_1 = \begin{array}{c|cccc} & d & e & f & g \\ \hline a & 1 & 2 & \perp & 3 \\ b & 4 & \perp & 5 & \perp \\ c & 6 & 7 & \perp & 8 \end{array}, \quad A_2 = \begin{array}{c|ccc} & d & i & f \\ \hline a & 11 & \perp & 12 \\ c & \perp & 13 & 14 \\ h & 15 & \perp & \perp \end{array}.$$

Then

$$A = \flat\{A_1, A_2\} = \begin{array}{c|ccccc} & d & e & f & g & i \\ \hline a & \langle 1, 11 \rangle & \langle 2, \perp \rangle & \langle \perp, 12 \rangle & \langle 3, \perp \rangle & \langle \perp, \perp \rangle \\ b & \langle 4, \perp \rangle & \langle \perp, \perp \rangle & \langle 5, \perp \rangle & \langle \perp, \perp \rangle & \langle \perp, \perp \rangle \\ c & \langle 6, \perp \rangle & \langle 7, \perp \rangle & \langle \perp, 14 \rangle & \langle 8, \perp \rangle & \langle \perp, 13 \rangle \\ h & \langle \perp, 15 \rangle & \langle \perp, \perp \rangle & \langle \perp, \perp \rangle & \langle \perp, \perp \rangle & \langle \perp, \perp \rangle \end{array}.$$

On the other hand,

$$\sharp(A) = \{@Pr_1(A), @Pr_2(A)\}$$

$$= \left\{ @ \begin{pmatrix} & d & e & f & g & i \\ \hline a & 1 & 2 & \perp & 3 & \perp \\ b & 4 & \perp & 5 & \perp & \perp \\ c & 6 & 7 & \perp & 8 & \perp \\ h & \perp & \perp & \perp & \perp & \perp \end{pmatrix}, @ \begin{pmatrix} & d & e & f & g & i \\ \hline a & 11 & \perp & 12 & \perp & \perp \\ b & \perp & \perp & \perp & \perp & \perp \\ c & \perp & \perp & 14 & \perp & 13 \\ h & 15 & \perp & \perp & \perp & \perp \end{pmatrix} \right\}$$

$$
= \left\{
\begin{array}{cc|cccc}
 & \begin{array}{cccc} d & e & f & g \end{array} & & \begin{array}{ccc} d & f & i \end{array} \\
\begin{array}{c} a \\ b \\ c \end{array} & \begin{array}{cccc} 1 & 2 & \bot & 3 \\ 4 & \bot & 5 & \bot \\ 6 & 7 & \bot & 8 \end{array} & , & \begin{array}{c} a \\ c \\ h \end{array} \begin{array}{ccc} 11 & 12 & \bot \\ \bot & 14 & 13 \\ 15 & \bot & \bot \end{array}
\end{array}
\right\}.
$$

Finally, we can define "**inflating operation**" that is defined for index sets $K \subset P \subset \mathcal{I}$ and $L \subset Q \subset \mathcal{I}$ by

$$
{}^{(P,Q)}A = {}^{(P,Q)}[K, L, \{a_{k_i,l_j}\}] = [P, Q, \{b_{p_r,q_s}\}],
$$

where

$$
b_{p_r,q_s} = \begin{cases} a_{k_i,l_j}, & \text{if } p_r = k_i \in K \text{ and } q_s = l_j \in L \\ \bot, & \text{otherwise} \end{cases}
$$

3.6 EIMs, Determinants and Permanents

As it is well-known, to each standard matrix can be juxtaposed a number, called determinant. Also, it is known that if we change the places of two rows or two columns of a standard matrix, the determinant of the new matrix will coincide with the former, but with an opposite sign (i.e., sign "+" is changed with sign "−" or vice versa). The same change of two rows or two columns of an IM, however, does not change the form of the new IM. For example, the two IMs from the above example satisfy the equality

$$
\begin{array}{c|ccc}
 & d & i & f \\
\hline
a & 11 & \bot & 12 \\
c & \bot & 13 & 14 \\
h & 15 & \bot & \bot
\end{array}
=
\begin{array}{c|ccc}
 & d & f & i \\
\hline
a & 11 & 12 & \bot \\
c & \bot & 14 & 13 \\
h & 15 & \bot & \bot
\end{array}.
$$

Of course, we can juxtapose a determinant only to a matrix with elements being real (complex) numbers. Having in mind the above equality, we can conclude that for IMs there is no possibility to juxtapose a determinant. On the other hand, to each IM with elements being real (or complex) numbers, we can juxtapose a permanent (see e.g., [42]).

Now, we extend this possibility to each EIM. For this reason, we must define some evaluating function $\Phi : \mathcal{X} \to \mathcal{R}$, such that it is an identity in the case $\mathcal{X} = \mathcal{R}$.

Let

$$
A = [K, L, \{a_{k_i,l_j}\}] =
\begin{array}{c|ccccc}
 & l_1 & \cdots & l_j & \cdots & l_n \\
\hline
k_1 & a_{k_1,l_1} & \cdots & a_{k_1,l_j} & \cdots & a_{k_1,l_n} \\
\vdots & \vdots & \cdots & \vdots & \cdots & \vdots \\
k_i & a_{k_i,l_1} & a_{k_i,l_j} & \cdots & a_{k_i,l_n} \\
\vdots & \vdots & \cdots & \vdots & \cdots & \vdots \\
k_m & a_{k_m,l_1} & \cdots & a_{k_m,l_j} & \cdots & a_{k_m,l_n}
\end{array},
$$

be an EIM for the elements of which there exists an evaluating function Φ. Then

$$per_\Phi(A) = \begin{cases} \displaystyle\sum_{[\rho(1),\ldots,\rho(m)]} \Phi(a)_{k_1,l_{\rho(1)}}\Phi(a)_{k_2,l_{\rho(2)}}\cdots\Phi(a)_{k_m,l_{\rho(m)}}, \\ \qquad\qquad \text{if } m = card(K) \leq card(L) = n \\ \\ \displaystyle\sum_{[\sigma(1),\ldots,\sigma(n)]} \Phi(a)_{k_{\sigma(1)},l_1}\Phi(a)_{k_{\sigma(2)},l_2}\cdots\Phi(a)_{k_{\sigma(n)},l_n}, \\ \qquad\qquad \text{if } m = card(K) \geq card(L) = n \end{cases},$$

where $\rho : \{1, 2, \ldots, m\} \rightarrow \{1, 2, \ldots, n\}$ is a bijection from $\{1, 2, \ldots, m\}$ in $\{1, 2, \ldots, n\}$ and $\sigma : \{1, 2, \ldots, n\} \rightarrow \{1, 2, \ldots, m\}$ is a bijection from $\{1, 2, \ldots, n\}$ in $\{1, 2, \ldots, m\}$.

And yet, we can keep the concept of a determinant adding additional condition. For brevity, here we discuss only the case of square EIM, i.e., the sets K and L have equal number of elements. Let us fix the order of the elements of set K as a vector $[k_1, \ldots, k_m]$. Each of its permutations $[k_{\rho(1)}, \ldots, k_{\rho(m)}]$ will be estimated as odd or even about the vector $[k_1, \ldots, k_m]$. Then, the determinant of the EIM A can be defined by

$$\begin{aligned} det_{\Phi,[k_1,\ldots,k_m]}(A) \\ = \sum_{[\rho(1),\ldots,\rho(m)]} (-1)^{[k_1,\ldots,k_m]}\Phi(a)_{k_1,l_{\rho(1)}}\Phi(a)_{k_2,l_{\rho(2)}}\cdots\Phi(a)_{k_m,l_{\rho(m)}}. \end{aligned}$$

3.7 Transposed EIM

Let the EIM A be given as above. Then its Transposed EIM has the form

$$A' = [L, K, \{a_{l_j,k_i}\}] = \begin{array}{c|ccccc} & k_1 & \cdots & k_i & \cdots & k_m \\ \hline l_1 & a_{l_1,k_1} & \cdots & a_{l_1,k_i} & \cdots & a_{l_1,k_m} \\ \vdots & \vdots & \cdots & \vdots & \cdots & \vdots \\ l_j & a_{l_j,k_1} & \cdots & a_{l_j,k_i} & \cdots & a_{l_j,k_n} \\ \vdots & \vdots & \cdots & \vdots & \cdots & \vdots \\ l_n & a_{l_n,k_1} & \cdots & a_{l_n,k_i} & \cdots & a_{l_n,k_m} \end{array}.$$

The geometrical interpretation is

$$\left(\begin{array}{c|cc} & L & \\ \hline K & \bullet & \circ \end{array}\right)' = \begin{array}{c|c} & K \\ \hline L & \bullet \\ & \circ \end{array}$$

The EIMs $A \odot_{(\circ,*)} A'$ and $A' \odot_{(\circ,*)} A$ are square EIMs. For example, if

$$A = \begin{array}{c|cc} & c & d \\ \hline a & \alpha & \beta \\ b & \gamma & \delta \\ e & \varepsilon & \zeta \end{array},$$

then

$$A' = \begin{array}{c|ccc} & a & b & c \\ \hline d & \alpha & \gamma & \varepsilon \\ e & \beta & \delta & \eta \end{array}$$

and

$$A \odot_{(+,\times)} A' = \begin{array}{c|ccc} & a & b & c \\ \hline a & \alpha^2 + \beta^2 & \alpha\gamma + \beta\delta & \alpha\varepsilon + \beta\zeta \\ b & \alpha\gamma + \beta\delta & \gamma^2 & \gamma\varepsilon + \delta\zeta \\ c & \alpha\varepsilon + \beta\zeta & \gamma\varepsilon + \delta\zeta & \varepsilon^2 + \zeta^2 \end{array},$$

while

$$A' \odot_{(+,\times)} A = \begin{array}{c|cc} & d & e \\ \hline d & \alpha^2 + \gamma^2 + \varepsilon^2 & \alpha\beta + \gamma\delta + \varepsilon\zeta \\ b & \alpha\beta + \gamma\delta + \varepsilon\zeta & \beta^2 + \delta^2 + \zeta^2 \end{array}.$$

3.8 An Example: An Intercriteria Decision Making Method

Following [22], the intercriteria decision making method, introduced by Deyan Mavrov, Vassia Atanassova and the author, is described.

Let us have an IM

$$A = \begin{array}{c|ccccccc} & O_1 & \cdots & O_k & \cdots & O_l & \cdots & O_n \\ \hline C_1 & a_{C_1,O_1} & \cdots & a_{C_1,O_k} & \cdots & a_{C_1,O_l} & \cdots & a_{C_1,O_n} \\ \vdots & \vdots & \vdots & \vdots & \vdots & \vdots & \vdots & \vdots \\ C_i & a_{C_i,O_1} & \cdots & a_{C_i,O_k} & \cdots & a_{C_i,O_l} & \cdots & a_{C_i,O_n} \\ \vdots & \vdots & \vdots & \vdots & \vdots & \vdots & \vdots & \vdots \\ C_j & a_{C_j,O_1} & \cdots & a_{C_j,O_k} & \cdots & a_{C_j,O_l} & \cdots & a_{C_j,O_n} \\ \vdots & \vdots & \vdots & \vdots & \vdots & \vdots & \vdots & \vdots \\ C_m & a_{C_m,O_1} & \cdots & a_{C_m,O_k} & \cdots & a_{C_m,O_l} & \cdots & a_{C_m,O_n} \end{array},$$

where for every p, q, $(1 \le p \le m, 1 \le q \le n)$:

- C_p is a criterion, taking part in the evaluation,
- O_q is an object, being evaluated.
- a_{C_p,O_q} is a real number or another object, that is comparable about relation R with the other a-objects, so that for each i, j, k: $R(a_{C_k,O_i}, a_{C_k,O_j})$ is defined. Let $\overline{R}$ be the dual relation of R in the sense that if R is satisfied, then $\overline{R}$ is not satisfied and vice versa. For example, if "R" is the relation "$<$", then $\overline{R}$ is the relation "$>$", and vice versa.

Let $S_{k,l}^{\mu}$ be the number of cases in which $R(a_{C_k,O_i}, a_{C_k,O_j})$ and $R(a_{C_l,O_i}, a_{C_l,O_j})$ are simultaneously satisfied. Let $S_{k,l}^{\nu}$ be the number of cases in which $R(a_{C_k,O_i}, a_{C_k,O_j})$ and $\overline{R}(a_{C_l,O_i}, a_{C_l,O_j})$ are simultaneously satisfied. Obviously,

$$S_{k,l}^{\mu} + S_{k,l}^{\nu} \leq \frac{n(n-1)}{2}.$$

Now, for every k, l, such that $1 \leq k < l \leq m$ and for $n \geq 2$, we define

$$\mu_{C_k,C_l} = 2\frac{S_{k,l}^{\mu}}{n(n-1)}, \quad \nu_{C_k,C_l} = 2\frac{S_{k,l}^{\nu}}{n(n-1)}.$$

Therefore, $\langle \mu_{C_k,C_l}, \nu_{C_k,C_l} \rangle$ is an IFP. Now, we can construct the IM

	C_1	$\cdots$	C_m
C_1	$\langle \mu_{C_1,C_1}, \nu_{C_1,C_1} \rangle$	$\cdots$	$\langle \mu_{C_1,C_m}, \nu_{C_1,C_m} \rangle$
$\vdots$	$\vdots$	$\vdots$	$\vdots$
C_m	$\langle \mu_{C_m,C_1}, \nu_{C_m,C_1} \rangle$	$\cdots$	$\langle \mu_{C_m,C_m}, \nu_{C_m,C_m} \rangle$

that determine the degrees of correspondence between criteria $C_1, \ldots, C_m$.

Let $\alpha, \beta \in [0, 1]$ be given, so that $\alpha + \beta \leq 1$. We say that criteria C_k and C_l are in

- (α, β)-positive consonance, if

$$\mu_{C_k,C_l} > \alpha \text{ and } \nu_{C_k,C_l} < \beta;$$

- (α, β)-negative consonance, if

$$\mu_{C_k,C_l} < \beta \text{ and } \mu_{C_k,C_l} > \alpha;$$

- (α, β)-dissonance, otherwise.

The method can be used for prediction.

Let the IM A be given and let criterion D (e.g., one of the criteria $C_1, \ldots, C_m$) be fixed. Let us reduce IM A to the IM

$$B = \begin{array}{c|ccccccc} & O_1 & \cdots & O_k & \cdots & O_l & \cdots & O_n \\ \hline C_1 & a_{C_1,O_1} & \cdots & a_{C_1,O_k} & \cdots & a_{C_1,O_l} & \cdots & a_{C_1,O_n} \\ \vdots & \vdots & \vdots & \vdots & \vdots & \vdots & \vdots & \vdots \\ C_i & a_{C_i,O_1} & \cdots & a_{C_i,O_k} & \cdots & a_{C_i,O_l} & \cdots & a_{C_i,O_n} \\ \vdots & \vdots & \vdots & \vdots & \vdots & \vdots & \vdots & \vdots \\ C_p & a_{C_p,O_1} & \cdots & a_{C_p,O_k} & \cdots & a_{C_p,O_l} & \cdots & a_{C_p,O_n} \\ D & b_{D,O_1} & \cdots & b_{D,O_k} & \cdots & b_{D,O_l} & \cdots & b_{D,O_n} \end{array}$$

omitting, if necessary, some rows, so that all criteria corresponding to the rows of B, be in (α, β)-positive or (α, β)-negative consonance with D.

For brevity, we say that these criteria are in consonance.

The important particularity in this case is that elements $b_{D,O_1}, \ldots, b_{D,O_n}$ are evaluated hardlier than the rest a-elements of B.

Let us have a new object X with estimations $x_1, \ldots, x_p$ w.r.t. the criteria $C_1, \ldots, C_p$. Then we can solve the following problem: "Predict the value y of object X w.r.t. criterion D".

To solve the problem, we can use one of the following two algorithms.

First Algorithm

We realize the following steps for each i, $1 \le i \le p$:

1.1. Determine the values a_{C_i,O_j} and a_{C_i,O_k} so that $a_{C_i,O_j} < a_{C_i,O_k}$ and $a_{C_i,O_j} \le x_i \le a_{C_i,O_k}$ and a_{C_i,O_j} is the highest a_{C_i,O_r} with this property and a_{C_i,O_k} is the lowest a_{C_i,O_s} with this property (for $1 \le r, s \le p$).

1.2. If criteria C_i and D are in positive consonance, then calculate the value

$$y_i = b_{D,O_j} + (x_i - a_{C_i,O_j}) \cdot \frac{b_{D,O_k} - b_{D,O_j}}{a_{C_i,O_k} - a_{C_i,O_j}}$$

and if criteria C_i and D are in negative consonance, then calculate the value

$$y_i = b_{D,O_j} + (x_i - a_{C_i,O_j}) \cdot \frac{b_{D,O_j} - b_{D,O_k}}{a_{C_i,O_k} - a_{C_i,O_j}}.$$

1.3. Determine the values

$$y_{min} = \min_{1 \le i \le p} y_i,$$

$$y_{ave} = \frac{1}{p} \sum_{1 \le i \le p} y_i,$$

$$y_{max} = \max_{1 \le i \le p} y_i.$$

Now, the value of y can be y_{ave} or some other number in interval $[y_{min}, y_{max}]$.

If there is no number a_{C_i,O_j} such that $a_{C_i,O_j} \le x_i$, or a_{C_i,O_k} such that $x_i \le a_{C_i,O_k}$, then Step 1.2 is omitted and in Step 1.3, the denominator is $p-s$, where s is the number of omitted cases (if they are smaller than p). If in Step 1.1, $a_{C_i,O_j} = x_i = a_{C_i,O_k}$ and $b_{D,O_j} < b_{D,O_k}$, then

$$y_i = \frac{1}{2}(b_{D,O_k} - b_{D,O_j})$$

for the case of positive consonance between criteria C_i and D and

$$y_i = \frac{1}{2}(b_{D,O_j} - b_{D,O_k})$$

for the case of negative consonance between these criteria.

Second Algorithm

2.1. Determine those objects O_j, for which for each i $(1 \leq i \leq p)$: $a_{C_i,O_j} \leq x_i$ and those objects O_k, for which for each i $(1 \leq i \leq p)$: $a_{C_i,O_k} \geq x_i$.
2.2. Determine object O_r, so that a_{C_i,O_r} is the highest a-element from the determined in Step 2.1 and $a_{C_i,O_r} \leq x_i$.
2.3. Determine object O_s, so that a_{C_i,O_s} is the lowest a-element from the determined in Step 2.1 and $a_{C_i,O_s} \geq x_i$.
2.4. Determine

$$y = \begin{cases} b_{D,O_r} + \frac{b_{D,O_s} - b_{D,O_r}}{p} \cdot \sum_{i=1}^{p} \frac{x_i - a_{C_i,O_r}}{a_{C_i,O_s} - a_{C_i,O_r}}, & \text{if } b_{D,O_s} \geq b_{D,O_r} \\[2em] b_{D,O_s} + \frac{b_{D,O_r} - b_{D,O_s}}{p} \cdot \sum_{i=1}^{p} \frac{x_i - a_{C_i,O_r}}{a_{C_i,O_s} - a_{C_i,O_r}}, & \text{otherwise} \end{cases}.$$

Now we discuss two (standard) formulas for evaluation of the y-values. Let the IM

$$B = \begin{array}{c|ccccc} & O_1 & \cdots & O_k & \cdots & O_n \\ \hline C_1 & a_{C_1,O_1} & \cdots & a_{C_1,O_k} & \cdots & a_{C_1,O_n} \\ \vdots & \vdots & \vdots & \vdots & \vdots & \vdots \\ C_i & a_{C_i,O_1} & \cdots & a_{C_i,O_k} & \cdots & a_{C_i,O_n} \\ \vdots & \vdots & \vdots & \vdots & \vdots & \vdots \\ C_p & a_{C_p,O_1} & \cdots & a_{C_p,O_k} & \cdots & a_{C_p,O_n} \\ D & b_{D,O_1} & \cdots & b_{D,O_k} & \cdots & b_{D,O_n} \end{array}$$

be given.

1. For every k $(1 \leq k \leq n)$ we construct the IM

$$B_k = B_{(\perp,O_k)}.$$

2. For every i $(1 \leq i \leq p)$ we put $x_i = a_{C_i,O_k}$.
3. Using the two above described methods (for the fixed number k), for B_k and $x_1, \ldots, x_p$, we determine y-values $y_{k,1}, y_{k,2}$.
4. For s $(s = 1, 2)$, we determine numbers

$$z_{k,s} = |y_{k,s} - b_{D,O_k}|.$$

5. Evaluate the standard deviation by:

$$\sigma_s' = \frac{1}{n(B_2 - B_1)} \sum_{k=1}^{n} z_{k,s},$$

$$\sigma_s'' = \frac{1}{B_2 - B_1} \sqrt{\frac{1}{n} \sum_{k=1}^{n} z_{k,s}^2}.$$

In future, the new method can be applied to different areas. For example, in medicine, it can show some intercriterial dependencies, related to criteria for decision making about the status of a patient from medical experts. The method can be used for searching of the values of objects, for which we have only partial information, and others.

We finish the Chapter with the following **Open problems**

4. Which is the IM- or EIM-interpretation of the tensors and which properties they will have?
5. Which is the form of IM- or EIM-determinant and which properties it will have?
6. Which other forms can have the hierarchical operator over EIMs?

Chapter 4
Temporal IFIMs

We introduce the definition of the object Temporal IFIM (TIFIM), described in the paper [25] of Evdokia Sotirova, Veselina Bureva, Anthony Shannon and the author, by

$$A(\mathcal{T}) = [K, L, \mathcal{T}, \{\langle \mu_{k_i,l_j,\tau}, \nu_{k_i,l_j,\tau} \rangle \}]$$

$$\equiv \left\{ \begin{array}{c|ccccc}
 & l_1 & l_2 & \cdots & l_n \\ \hline
k_1 & \langle \mu_{k_1,l_1,\tau}, \nu_{k_1,l_1,\tau} \rangle & \langle \mu_{k_1,l_2,\tau}, \nu_{k_1,l_2,\tau} \rangle & \cdots & \langle \mu_{k_1,l_n,\tau}, \nu_{k_1,l_n,\tau} \rangle \\
\vdots & \vdots & \vdots & \cdots & \vdots \\
k_i & \langle \mu_{k_i,l_1,\tau}, \nu_{k_i,l_1,\tau} \rangle & \langle \mu_{k_i,l_2,\tau}, \nu_{k_i,l_2,\tau} \rangle & \cdots & \langle \mu_{k_i,l_n,\tau}, \nu_{k_i,l_n,\tau} \rangle \\
\vdots & \vdots & \vdots & \cdots & \vdots \\
k_m & \langle \mu_{k_m,l_1,\tau}, \nu_{k_m,l_1,\tau} \rangle & \langle \mu_{k_m,l_2,\tau}, \nu_{k_m,l_2,\tau} \rangle & \cdots & \langle \mu_{k_m,l_n,\tau}, \nu_{k_m,l_n,\tau} \rangle
\end{array} \middle| \tau \in \mathcal{T} \right\},$$

where for every $\tau \in \mathcal{T}$, $1 \le i \le m$, $1 \le j \le n$:

$$\mu_{k_i,l_j,\tau}, \nu_{k_i,l_j,\tau}, \mu_{k_i,l_j,\tau} + \nu_{k_i,l_j,\tau} \in [0, 1].$$

Here, $\mathcal{T}$ is a some fixed temporal scale and τ is its element, i.e., a time-moment.

By analogy with the Chap. 2, we extend the concept of a TIFIM, defining an Extended TIFIM (ETIFIM; see [20]), by:

$$A^*(\mathcal{T}) = [K^*(\mathcal{T}), L^*(\mathcal{T}), \{\langle \mu_{k_i,l_j,\tau}, \nu_{k_i,l_j,\tau} \rangle \}]$$

© Springer International Publishing Switzerland 2014
K.T. Atanassov, *Index Matrices: Towards an Augmented Matrix Calculus*,
Studies in Computational Intelligence 573, DOI 10.1007/978-3-319-10945-9_4

$$\equiv \left\{ \begin{array}{c|ccc} & l_1, \langle \alpha^l_{1,\tau}, \beta^l_{1,\tau} \rangle & \cdots & l_n, \langle \alpha^l_{n,\tau}, \beta^l_{n,\tau} \rangle \\ \hline k_1, \langle \alpha^k_{1,\tau}, \beta^k_{1,\tau} \rangle & \langle \mu_{k_1,l_1,\tau}, \nu_{k_1,l_1,\tau} \rangle & \cdots & \langle \mu_{k_1,l_n,\tau}, \nu_{k_1,l_n,\tau} \rangle \\ \vdots & \vdots & \cdots & \vdots \\ k_i, \langle \alpha^k_{i,\tau}, \beta^k_{i,\tau} \rangle & \langle \mu_{k_i,l_1,\tau}, \nu_{k_i,l_1,\tau} \rangle & \cdots & \langle \mu_{k_i,l_n,\tau}, \nu_{k_i,l_n,\tau} \rangle \\ \vdots & \vdots & \cdots & \vdots \\ k_m, \langle \alpha^k_{m,\tau}, \beta^k_{m,\tau} \rangle & \langle \mu_{k_m,l_1,\tau}, \nu_{k_m,l_1,\tau} \rangle & \cdots & \langle \mu_{k_m,l_n,\tau}, \nu_{k_m,l_n,\tau} \rangle \end{array} \middle| \tau \in T \right\},$$

where for every $1 \le i \le m, 1 \le j \le n$:

$$\mu_{k_i,l_j,\tau}, \nu_{k_i,l_j,\tau}, \mu_{k_i,l_j,\tau} + \nu_{k_i,l_j,\tau} \in [0, 1],$$

$$\alpha^k_{i,\tau}, \beta^k_{i,\tau}, \alpha^k_{i,\tau} + \beta^k_{i,\tau} \in [0, 1],$$

$$\alpha^l_{j,\tau}, \beta^l_{j,\tau}, \alpha^l_{j,\tau} + \beta^l_{j,\tau} \in [0, 1]$$

and

$$K^*(T) = \{\langle k_i, \alpha^k_{i,\tau}, \beta^k_{i,\tau} \rangle | k_i \in K \ \& \ \tau \in T\}$$

$$= \{\langle k_i, \alpha^k_{i,\tau}, \beta^k_{i,\tau} \rangle | 1 \le i \le m \ \& \ \tau \in T\},$$

$$L^*(T) = \{\langle l_j, \alpha^l_{j,\tau}, \beta^l_{j,\tau} \rangle | l_j \in L \ \& \ \tau \in T\}$$

$$= \{\langle l_j, \alpha^l_{j,\tau}, \beta^l_{j,\tau} \rangle | 1 \le j \le n \ \& \ \tau \in T\}.$$

Let

$$K^*(T) \subset P^*(T) \text{ iff } (K \subset P) \ \& \ (\forall \tau \in T)$$

$$(\forall k_i = p_i \in K : (\alpha^k_{i,\tau} < \alpha^p_{i,\tau}) \ \& \ (\beta^k_{i,\tau} > \beta^p_{i,\tau})).$$

$$K^*(T) \subseteq P^*(T) \text{ iff } (K \subseteq P) \ \& \ (\forall \tau \in T)$$

$$(\forall k_i = p_i \in K : (\alpha^k_{i,\tau} \le \alpha^p_{i,\tau}) \ \& \ (\beta^k_{i,\tau} \ge \beta^p_{i,\tau})).$$

We must mention that the new IM is 3-dimensional by analogy with the IM described in Chap. 6. The indices for the third dimension of the TIFIM and ETIFIM are elements of a time-scale T:

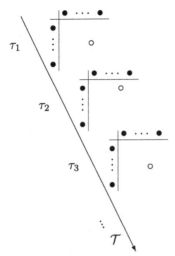

As above, here we discuss the operations, relations and operators over the extended type of TIFIMs.

4.1 Operations Over ETIFIMs

For the ETIFIMs

$$A^*(\mathcal{T}) = [K^*(\mathcal{T}), L^*(\mathcal{T}), \{\langle \mu_{k_i,l_j,\tau}, \nu_{k_i,l_j,\tau}\rangle\}],$$

$$B^*(\mathcal{T}) = [P^*(\mathcal{T}), Q^*(\mathcal{T}), \{\langle \rho_{p_r,q_s,\tau}, \sigma_{p_r,q_s,\tau}\rangle\}],$$

and for $(\circ, *) \in \{(\max, \min), (\min, \max)\}$, operations are the following.

Addition

$$A^*(\mathcal{T}) \oplus_{(\circ,*)} B^*(\mathcal{T}) = [T^*(\mathcal{T}), V^*(\mathcal{T}), \{\langle \varphi_{t_u,v_w,\tau}, \psi_{t_u,v_w,\tau}\rangle\}],$$

where

$$T^*(\mathcal{T}) = K^*(\mathcal{T}) \cup P^*(\mathcal{T}) = \{\langle t_u, \alpha^t_{u,\tau}, \beta^t_{u,\tau}\rangle | t_u \in K \cup P \,\&\, \tau \in T\},$$

$$V^*(\mathcal{T}) = L^*(\mathcal{T}) \cup Q^*(\mathcal{T}) = \{\langle v_w, \alpha^v_{w,\tau}, \beta^v_{w,\tau}\rangle | v_w \in L \cup Q \,\&\, \tau \in T\},$$

$$\alpha^t_{u,\tau} = \begin{cases} \alpha^k_{i,\tau}, & \text{if } t_u \in K - P \\ \alpha^p_{r,\tau}, & \text{if } t_u \in P - K, \\ \max(\alpha^k_{i,\tau}, \alpha^p_{r,\tau}), & \text{if } t_u \in K \cap P \end{cases}$$

$$\beta^v_{w,\tau} = \begin{cases} \beta^l_{j,\tau}, & \text{if } v_w \in L - Q \\ \beta^q_{s,\tau}, & \text{if } t_w \in Q - L, \\ \min(\beta^l_{j,\tau}, \beta^q_{s,\tau}), & \text{if } t_w \in L \cap Q \end{cases}$$

and

$$\langle \varphi_{t_u,v_w,\tau}, \psi_{t_u,v_w,\tau} \rangle =$$

$$= \begin{cases} \langle \mu_{k_i,l_j,\tau}, \nu_{k_i,l_j,\tau} \rangle, & \text{if } t_u = k_i \in K \text{ and } v_w = l_j \in L - Q \\ & \text{or } t_u = k_i \in K - P \text{ and } v_w = l_j \in L; \\ \langle \rho_{p_r,q_s,\tau}, \sigma_{p_r,q_s,\tau} \rangle, & \text{if } t_u = p_r \in P \text{ and } v_w = q_s \in Q - L \\ & \text{or } t_u = p_r \in P - K \text{ and } v_w = q_s \in Q;, \\ \langle \circ(\mu_{k_i,l_j,\tau}, \rho_{p_r,q_s,\tau}), & \text{if } t_u = k_i = p_r \in K \cap P \\ *(\nu_{k_i,l_j,t}, \sigma_{p_r,q_s,t}) \rangle, & \text{and } v_w = l_j = q_s \in L \cap Q \\ \langle 0, 1 \rangle, & \text{otherwise} \end{cases}$$

Termwise multiplication

$$A^*(T) \otimes_{(\circ,*)} B^*(T) = [T^*(T), V^*(T), \{\langle \varphi_{t_u,v_w,\tau}, \psi_{t_u,v_w,\tau} \rangle\}],$$

where

$$T^*(T) = K^*(T) \cap P^*(T) = \{\langle t_u, \alpha^t_{u,\tau}, \beta^t_{u,\tau} \rangle | t_u \in K \cap P \ \& \ \tau \in T\},$$

$$V^*(T) = L^*(T) \cap Q^*(T) = \{\langle v_w, \alpha^v_{w,\tau}, \beta^v_{w,\tau} \rangle | v_w \in L \cap Q \ \& \ \tau \in T\},$$

$$\alpha^t_{u,\tau} = \min(\alpha^k_{i,\tau}, \alpha^P_{r,\tau}), \text{ for } t_u = k_i = p_r \in K \cap P,$$

$$\beta^v_{w,\tau} = \max(\beta^l_{j,\tau}, \beta^q_{s,\tau}), \text{ for } v_w = l_j = q_s \in L \cap Q$$

and

$$\langle \varphi_{t_u,v_w,\tau}, \psi_{t_u,v_w,\tau} \rangle = \langle \circ(\mu_{k_i,l_j,\tau}, \rho_{p_r,q_s,\tau}), *(\nu_{k_i,l_j,\tau}, \sigma_{p_r,q_s,\tau}) \rangle.$$

Multiplication

$$A^*(T) \odot_{(\circ,*)} B^*(T) = [T^*(T), V^*(T), \{\langle \varphi_{t_u,v_w,\tau}, \psi_{t_u,v_w,\tau} \rangle\}],$$

where

$$T^*(T) = (K \cup (P - L))^*(T) = \{\langle t_u, \alpha^t_{u,\tau}, \beta^t_{u,\tau} \rangle | t_u \in K \cup (P - L)\},$$

$$V^*(T) = (Q \cup (L - P))^*(T) = \{\langle v_w, \alpha^v_{w,\tau}, \beta^v_{w,\tau} \rangle | v_w \in Q \cup (L - P)\},$$

$$\alpha^t_{u,\tau} = \begin{cases} \alpha^k_{i,\tau}, & \text{if } t_u = k_i \in K \\ \alpha^p_{r,\tau}, & \text{if } t_u = p_r \in P - L \end{cases},$$

$$\beta^v_{w,\tau} = \begin{cases} \beta^l_{j,\tau}, & \text{if } v_w = l_j \in L - P \\ \beta^q_{s,\tau}, & \text{if } t_w = q_s \in Q \end{cases},$$

and

$$\langle \varphi_{t_u,v_w,\tau}, \psi_{t_u,v_w,\tau} \rangle =$$

$$= \begin{cases} \langle \mu_{k_i,l_j,\tau}, \nu_{k_i,l_j,\tau} \rangle, & \begin{array}{l} \text{if } t_u = k_i \in K \\ \text{and } v_w = l_j \in L - P - Q \end{array} \\ \langle \rho_{p_r,q_s,\tau}, \sigma_{p_r,q_s,\tau} \rangle, & \begin{array}{l} \text{if } t_u = p_r \in P - L - K \\ \text{and } v_w = q_s \in Q \end{array} \\ \langle \underset{l_j=p_r \in L \cap P}{\circ} (\min(\mu_{k_i,l_j,\tau}, \rho_{p_r,q_s,\tau})), & \\ \quad \underset{l_j=p_r \in L \cap P}{*} (\max(\nu_{k_i,l_j,\tau}, \sigma_{p_r,q_s,\tau})) \rangle, & \text{if } t_u = k_i \in K \text{ and } v_w = q_s \in Q \\ \langle 0, 1 \rangle, & \text{otherwise} \end{cases}$$

Structural subtraction

$$A^*(\mathcal{T}) \ominus B^*(\mathcal{T}) = [T^*(\mathcal{T}), V^*(\mathcal{T}), \{\langle \varphi_{t_u,v_w,\tau}, \psi_{t_u,v_w,\tau} \rangle\}],$$

where

$$T^*(\mathcal{T}) = (K - P)^*(\mathcal{T}) = \{\langle t_u, \alpha^t_{u,\tau}, \beta^t_{u,\tau} \rangle | t_u \in K - P\},$$

$$V^*(\mathcal{T}) = (L - Q)^*(\mathcal{T}) = \{\langle v_w, \alpha^v_{w,\tau}, \beta^v_{w,\tau} \rangle | v_w \in L - Q\},$$

for the set–theoretic subtraction operation and

$$\alpha^t_{u,\tau} = \alpha^k_{i,\tau}, \text{ for } t_u = k_i \in K - P,$$

$$\beta^v_{w,\tau} = \beta^l_{j,\tau}, \text{ for } v_w = l_j \in L - Q$$

and

$$\langle \varphi_{t_u,v_w,\tau}, \psi_{t_u,v_w,\tau} \rangle = \langle \mu_{k_i,l_j,\tau}, \nu_{k_i,l_j,\tau} \rangle,$$

for $t_u = k_i \in K - P$ and $v_w = l_j \in L - Q$.

Negation of an ETIFIM

$$\neg A(\mathcal{T})^* = [T^*(\mathcal{T}), V^*(\mathcal{T}), \{\neg \langle \mu_{k_i,l_j,\tau}, \nu_{k_i,l_j,\tau} \rangle\}],$$

where $\neg$ is one of the negations from Table 2.1 from Sect. 2.1, or another possibly defined.

Termwise subtraction

$$A^*(T) -_{(\circ,*)} B^*(T) = A^*(T) \oplus_{(\circ,*)} \neg B^*(T).$$

Operations "reduction", "projection" and "substitution" coincide with the respective operations defined over IMs in [10], while the hierarchical operations over IMs are not applicable here.

4.2 Relations Over ETIFIMs

Let the two ETIFIMs

$$A^*(T) = [K^*(T), L^*(T), \{\langle \mu_{k_i,l_j,\tau}, \nu_{k_i,l_j,\tau}\rangle\}]$$

and

$$B^*(T) = [P^*(T), Q^*(T), \{\langle \rho_{p_r,q_s,\tau}, \sigma_{p_r,q_s,\tau}\rangle\}]$$

be given. We introduce the following definitions where $\subset$ and $\subseteq$ denote the relations "*strong inclusion*" and "*weak inclusion*".

The strict relation "inclusion about matrix-dimension and elements" is

$$A^*(T) \subset_d^e B^*(T) \text{ iff } (((K^*(T) \subset P^*(T)) \& (L^*(T) \subset Q^*(T)))$$

$$\vee ((K^*(T) \subseteq P^*(T)) \& (L^*(T) \subset Q^*(T)))$$

$$\vee (K^*(T) \subset P^*(T)) \& (L^*(T) \subseteq Q^*(T)))$$

$$\& (\forall \tau \in (T))(\forall k \in K)(\forall l \in L)(\langle \mu_{k_i,l_j,\tau}, \nu_{k_i,l_j,\tau}\rangle = \langle \rho_{k_i,l_j,\tau}, \sigma_{k_i,l_j,\tau}\rangle).$$

The non-strict relation "inclusion about matrix-dimension and elements" is

$$A^*(T) \subseteq_d^e B^*(T) \text{ iff } (K^*(T) \subseteq P^*(T)) \& (L^*(T) \subseteq Q^*(T))$$

$$\& (\forall \tau \in (T))(\forall k \in K)(\forall l \in L)(\langle \mu_{k_i,l_j,\tau}, \nu_{k_i,l_j,\tau}\rangle = \langle \rho_{k_i,l_j,\tau}, \sigma_{k_i,l_j,\tau}\rangle).$$

The strict relation "inclusion about element values" is

$$A^*(T) \subset_v^e B^*(T) \text{ iff } (K^*(T) = P^*(T)) \& (L^*(T) = Q^*(T))$$

$$\& (\forall \tau \in (T))(\forall k \in K)(\forall l \in L)(\langle \mu_{k_i,l_j,\tau}, \nu_{k_i,l_j,\tau}\rangle < \langle \rho_{k_i,l_j,\tau}, \sigma_{k_i,l_j,\tau}\rangle).$$

The non-strict relation "inclusion about element values" is

$$A^*(T) \subseteq_v^e B^*(T) \text{ iff } (K^*(T) = P^*(T)) \& (L^*(T) = Q^*(T))$$

$$\& (\forall \tau \in (T))(\forall k \in K)(\forall l \in L)(\langle \mu_{k_i, l_j, \tau}, \nu_{k_i, l_j, \tau} \rangle \le \langle \rho_{k_i, l_j, \tau}, \sigma_{k_i, l_j, \tau} \rangle).$$

The strict relation "inclusion about matrix-dimension and element values" is

$$A^*(T) \subset^e B^*(T) \text{ iff } (((K^*(T) \subset P^*(T)) \& (L^*(T) \subset Q^*(T)))$$

$$\vee ((K^*(T) \subseteq P^*(T)) \& (L^*(T) \subset Q^*(T)))$$

$$\vee ((K^*(T) \subset P^*(T)) \& (L^*(T) \subseteq Q)^*(T)))$$

$$\& (\forall \tau \in (T))(\forall k \in K)(\forall l \in L)(\langle \mu_{k_i, l_j, \tau}, \nu_{k_i, l_j, \tau} \rangle < \langle \rho_{k_i, l_j, \tau}, \sigma_{k_i, l_j, \tau} \rangle).$$

The non-strict relation "inclusion about matrix-dimension and element values"
is

$$A^*(T) \subseteq^e B^*(T) \text{ iff } (K^*(T) \subseteq P^*(T)) \& (L^*(T) \subseteq Q^*(T))$$

$$\& (\forall \tau \in (T))(\forall k \in K)(\forall l \in L)(\langle \mu_{k_i, l_j, \tau}, \nu_{k_i, l_j, \tau} \rangle \le \langle \rho_{k_i, l_j, \tau}, \sigma_{k_i, l_j, \tau} \rangle).$$

The strict relation "inclusion about matrix-dimension and indices" is

$$A^*(T) \subset_d^i B^*(T) \text{ iff } (((K^*(T) \subset P^*(T)) \& (L^*(T) \subset Q^*(T)))$$

$$\vee ((K^*(T) \subseteq P^*(T)) \& (L^*(T) \subset Q^*(T)))$$

$$\vee ((K^*(T) \subset P^*(T)) \& (L^*(T) \subseteq Q^*(T))))\& (\forall \tau \in (T))(\forall k \in K)(\forall l \in L)$$

$$(\langle \alpha_{i,\tau}^k, \beta_{i,\tau}^k \rangle = \langle \alpha_{i,\tau}^p, \beta_{i,\tau}^p \rangle \& \langle \alpha_{i,\tau}^l, \beta_{i,\tau}^l \rangle = \langle \alpha_{i,\tau}^q, \beta_{i,\tau}^q \rangle).$$

The non-strict relation "inclusion about matrix-dimension and indices" is

$$A^*(T) \subseteq_d^i B^*(T) \text{ iff } (K^*(T) \subseteq P^*(T)) \& (L^*(T) \subseteq Q^*(T))$$

$$\& (\forall \tau \in (T))(\forall k \in K)(\forall l \in L)$$

$$(\langle \alpha_{i,\tau}^k, \beta_{i,\tau}^k \rangle = \langle \alpha_{i,\tau}^p, \beta_{i,\tau}^p \rangle \& \langle \alpha_{i,\tau}^l, \beta_{i,\tau}^l \rangle = \langle \alpha_{i,\tau}^q, \beta_{i,\tau}^q \rangle).$$

The strict relation "inclusion about matrix-dimension and index values" is

$$A^*(T) \subset_v^i B^*(T) \text{ iff } (K^*(T) = P^*(T)) \& (L^*(T) = Q^*(T))$$

$$\& (\forall \tau \in (T))(\forall k \in K)(\forall l \in L)$$

$$(\langle \alpha^k_{i,\tau}, \beta^k_{i,\tau} \rangle < \langle \alpha^p_{i,\tau}, \beta^p_{i,\tau} \rangle \,\&\, \langle \alpha^l_{i,\tau}, \beta^l_{i,\tau} \rangle < \langle \alpha^q_{i,\tau}, \beta^q_{i,\tau} \rangle).$$

The non-strict relation "inclusion about matrix-dimension and index values" is

$$A^*(T) \subseteq^i_v B^*(T) \text{ iff } (K^*(T) = P^*(T)) \,\&\, (L^*(T) = Q^*(T))$$

$$\&\; (\forall \tau \in (T))(\forall k \in K)(\forall l \in L)$$

$$(\langle \alpha^k_{i,\tau}, \beta^k_{i,\tau} \rangle \le \langle \alpha^p_{i,\tau}, \beta^p_{i,\tau} \rangle \,\&\, \langle \alpha^l_{i,\tau}, \beta^l_{i,\tau} \rangle \le \langle \alpha^q_{i,\tau}, \beta^q_{i,\tau} \rangle).$$

The strict relation "inclusion about index values" is

$$A^*(T) \subset^i B^*(T) \text{ iff } (((K^*(T) \subset P^*(T)) \,\&\, (L^*(T) \subset Q^*(T)))$$

$$\vee\, ((K^*(T) \subseteq P^*(T)) \,\&\, (L^*(T) \subset Q^*(T)))$$

$$\vee\, ((K^*(T) \subset P^*(T)) \&(L^*(T) \subseteq Q)^*(T))) \,\&\, (\forall \tau \in (T))(\forall k \in K)(\forall l \in L)$$

$$(\langle \alpha^k_{i,\tau}, \beta^k_{i,\tau} \rangle < \langle \alpha^p_{i,\tau}, \beta^p_{i,\tau} \rangle \,\&\, \langle \alpha^l_{i,\tau}, \beta^l_{i,\tau} \rangle < \langle \alpha^q_{i,\tau}, \beta^q_{i,\tau} \rangle).$$

The non-strict relation "inclusion about index values" is

$$A^*(T) \subseteq^i B^*(T) \text{ iff } (K^*(T) \subseteq P^*(T)) \,\&\, (L^*(T) \subseteq Q^*(T))$$

$$\&\; (\forall \tau \in (T))(\forall k \in K)(\forall l \in L)$$

$$(\langle \alpha^k_{i,\tau}, \beta^k_{i,\tau} \rangle \le \langle \alpha^p_{i,\tau}, \beta^p_{i,\tau} \rangle \,\&\, \langle \alpha^l_{i,\tau}, \beta^l_{i,\tau} \rangle \le \langle \alpha^q_{i,\tau}, \beta^q_{i,\tau} \rangle).$$

ETIFIM $A^*(T)$ **has temporal strictly increasing elements**, if

$$(\forall \tau_1, \tau_2 \in T)((\tau_1 < \tau_2) \to (\forall k_i \in K)(\forall l_j \in L)$$

$$(\langle \mu_{k_i,l_j,\tau_1}, \nu_{k_i,l_j,\tau_1} \rangle < \langle \rho_{k_i,l_j,\tau_2}, \sigma_{k_i,l_j,\tau_2} \rangle).$$

ETIFIM $A^*(T)$ **has temporal non-strictly increasing elements**, if

$$(\forall \tau_1, \tau_2 \in T)((\tau_1 < \tau_2) \to (\forall k_i \in K)(\forall l_j \in L)$$

$$(\langle \mu_{k_i,l_j,\tau_1}, \nu_{k_i,l_j,\tau_1} \rangle \le \langle \rho_{k_i,l_j,\tau_2}, \sigma_{k_i,l_j,\tau_2} \rangle).$$

ETIFIM $A^*(T)$ **has temporal strictly increasing indices**, if

$$(\forall \tau_1, \tau_2 \in T)((\tau_1 < \tau_2) \to (\forall k_i \in K)(\forall l_j \in L)$$

$$(\langle \alpha^k_{i,\tau_1}, \beta^k_{i,\tau_1}\rangle < \langle \alpha^p_{i,\tau_2}, \beta^p_{i,\tau_2}\rangle \ \& \ \langle \alpha^l_{j,\tau_1}, \beta^l_{j,\tau_1}\rangle < \langle \alpha^q_{j,\tau_2}, \beta^q_{j,\tau_2}\rangle).$$

ETIFIM $A^*(\mathcal{T})$ has temporal non-strictly increasing indices, if

$$(\forall \tau_1, \tau_2 \in \mathcal{T})((\tau_1 < \tau_2) \to (\forall k_i \in K)(\forall l_j \in L)$$

$$(\langle \alpha^k_{i,\tau_1}, \beta^k_{i,\tau_1}\rangle \leq \langle \alpha^p_{i,\tau_2}, \beta^p_{i,\tau_2}\rangle \ \& \ \langle \alpha^l_{j,\tau_1}, \beta^l_{j,\tau_1}\rangle \leq \langle \alpha^q_{j,\tau_2}, \beta^q_{j,\tau_2}\rangle).$$

4.3 Specific Operations Over ETIFIMs

Let operations $\bar{\circ}$ and $\bar{*}$ are dual operations of operations $\circ$ and $*$, respectively. For example, pairs $(\circ, \bar{\circ})$ and $(*, \bar{*})$ can be any among pairs (max, min), (min, max). If $\circ$ and $*$ are average operations, then $\bar{\circ}$ and $\bar{*}$ are also average operations.

Let the time-scale $\mathcal{T}$ be fixed, let the ETIFIM

$$A^*(\mathcal{T}) = [K^*(\mathcal{T}), L^*(\mathcal{T}), \{\langle \mu_{k_i,l_j,\tau}, \nu_{k_i,l_j,\tau}\rangle\}]$$

$$\equiv \left\{ \begin{array}{c|ccc} & l_1, \langle \alpha^l_{1,\tau}, \beta^l_{1,\tau}\rangle & \cdots & l_n, \langle \alpha^l_{n,\tau}, \beta^l_{n,\tau}\rangle \\ \hline k_1, \langle \alpha^k_{1,\tau}, \beta^k_{1,\tau}\rangle & \langle \mu_{k_1,l_1,\tau}, \nu_{k_1,l_1,\tau}\rangle & \cdots & \langle \mu_{k_1,l_n,\tau}, \nu_{k_1,l_n,\tau}\rangle \\ \vdots & \vdots & \cdots & \vdots \\ k_m, \langle \alpha^k_{m,\tau}, \beta^k_{m,\tau}\rangle & \langle \mu_{k_m,l_1,\tau}, \nu_{k_m,l_1,\tau}\rangle & \cdots & \langle \mu_{k_m,l_n,\tau}, \nu_{k_m,l_n,\tau}\rangle \end{array} \middle| \tau \in \mathcal{T} \right\}$$

be given and let $k_0 \notin K$ and $l_0 \notin L$ be two indices. Now, we introduce the following operations over it:

$(\circ, *)$-**row-aggregation**

$$\rho_{(\circ,*)}(A^*(\mathcal{T}), k_0)$$

$$= \left\{ \begin{array}{c|c} & l_1, \langle \alpha^l_{1,\tau}, \beta^l_{1,\tau}\rangle \qquad \cdots \\ \hline k_0, \langle \underset{1\leq i \leq m}{\circ} \alpha^k_{i,\tau}, \underset{1\leq i \leq m}{\bar{\circ}} \beta^k_{i,\tau}\rangle & \langle \underset{1\leq i \leq m}{*} \mu_{k_i,l_1,\tau}, \underset{1\leq i \leq m}{\bar{*}} \nu_{k_i,l_1,\tau}\rangle \cdots \end{array} \right.$$

$$\left. \begin{array}{c} \cdots \qquad l_n, \langle \alpha^l_{n,\tau}, \beta^l_{n,\tau}\rangle \\ \hline \cdots \langle \underset{1\leq i \leq m}{*} \mu_{k_i,l_n,\tau}, \underset{1\leq i \leq m}{\bar{*}} \nu_{k_i,l_n,\tau}\rangle \end{array} \middle| \tau \in \mathcal{T} \right\},$$

$(\circ, *)$-**column-aggregation**

$$\sigma_{(\circ,*)}(A^*(\mathcal{T}), l_0)$$

$$
= \left\{
\begin{array}{c|c}
 & l_0, \langle\; \underset{1\le j\le n}{\circ}\; \alpha^l_{j,\tau},\; \underset{1\le j\le n}{\overline{\circ}}\; \beta^l_{j,\tau} \rangle \\[2ex]
\hline
k_1, \langle \alpha^k_{1,\tau}, \beta^k_{1,\tau} \rangle & \langle\; \underset{1\le j\le n}{*}\; \mu_{k_1,l_j,\tau},\; \underset{1\le j\le n}{\overline{*}}\; \nu_{k_i,l_j,\tau} \rangle \\[2ex]
\vdots & \vdots \\[2ex]
k_m, \langle \alpha^k_{m,\tau}, \beta^k_{m,\tau} \rangle & \langle\; \underset{1\le j\le n}{*}\; \mu_{k_m,l_j,\tau},\; \underset{1\le j\le n}{\overline{*}}\; \nu_{k_i,l_j,\tau} \rangle
\end{array}
\; \Big|\; \tau \in \mathcal{T} \right\}.
$$

Let $\bullet$ and $\overline{\bullet}$ be two other dual operations, that can coincide with $(\circ, \overline{\circ})$ or $(*, \overline{*})$, or to be different. The following new aggregation operation can be defined:

$(\circ, \bullet, *) - \mathcal{T}$-**aggregation**

$$
\rho^{\mathcal{T}}_{(\circ,\bullet,*)}(A^*(\mathcal{T}))
$$

$$
= \begin{array}{c|c}
 & l_1, \langle\; \underset{\tau \in \mathcal{T}}{\bullet}\; \alpha^l_{1,\tau},\; \underset{\tau \in \mathcal{T}}{\overline{\bullet}}\; \beta^l_{1,\tau} \rangle\;\; \cdots \\[2ex]
\hline
k_1, \langle\; \underset{\tau \in \mathcal{T}}{\circ}\; \alpha^k_{1,\tau},\; \underset{\tau \in \mathcal{T}}{\overline{\circ}}\; \beta^k_{1,\tau} \rangle & \langle\; \underset{\tau \in \mathcal{T}}{*}\; \mu_{k_1,l_1,\tau},\; \underset{\tau \in \mathcal{T}}{\overline{*}}\; \nu_{k_1,l_1,\tau} \rangle\;\; \cdots \\[2ex]
\vdots & \vdots \qquad \cdots \\[2ex]
k_m, \langle\; \underset{\tau \in \mathcal{T}}{\circ}\; \alpha^k_{m,\tau},\; \underset{\tau \in \mathcal{T}}{\overline{\circ}}\; \beta^k_{m,\tau} \rangle & \langle\; \underset{\tau \in \mathcal{T}}{*}\; \mu_{k_m,l_1,\tau},\; \underset{\tau \in \mathcal{T}}{\overline{*}}\; \nu_{k_m,l_1,\tau} \rangle\;\; \cdots
\end{array}
$$

$$
\begin{array}{c}
\cdots \quad l_n, \langle \alpha^l_{n,\tau}, \beta^l_{n,\tau} \rangle \\[2ex]
\hline
\cdots \quad \langle\; \underset{\tau \in \mathcal{T}}{*}\; \mu_{k_1,l_n,\tau},\; \underset{\tau \in \mathcal{T}}{\overline{*}}\; \nu_{k_1,l_n,\tau} \rangle \\[2ex]
\vdots \qquad \vdots \\[2ex]
\cdots \quad \langle\; \underset{\tau \in \mathcal{T}}{*}\; \mu_{k_m,l_n,\tau},\; \underset{\tau \in \mathcal{T}}{\overline{*}}\; \nu_{k_m,l_n,\tau} \rangle\rangle
\end{array}
$$

The above operations are extensions of the operations from IFIMs, EIFIMs and TIFIMs and have similar properties.

4.4 An Example: Temporal Intuitionistic Fuzzy Cognitive Map

In [27] Axelrod introduced the concept of a cognitive map as a mathematical and graphical representation of a persons system of beliefs. Cognitive maps, also called causal maps, are built up of only two fundamental elements, i.e. concepts (or factors) that represent the variables in the analyzed system and causal beliefs (or relationships) that determine the causal relations among those variables. Ten years later, in 1986, starting from the generic model of Axelrods cognitive maps, Kosko [40] introduced fuzzy cognitive maps as an extension of the latter.

Now, there are several possible formal definitions of FCMs in the literature. Probably the most commonly used formal definition is in the form given by Chen [30], which respects the original numerical matrix representation proposed by Kosko, where FCM is defined as a 4-tuple,

$$CM = (C, E, \alpha, \beta),$$

where:

- C is finite set of cognitive units (i.e., concepts), $C = \{C_1, C_2, \cdot, C_n\}$;
- E is a finite set of directed edges between cognitive units, $E = \{e_1, e_2, \cdot, e_m\}$;
- α is a mapping function from cognitive units to an interval $[a, b]$, where $-1 \leq a \leq b \leq +1$;
- $\beta : E \to [-1, +1]$ is a mapping function from directed edges to real values between -1 and $+1$.

Next step of the development of this concept is made by De et al. [32], by Iakovidis, and Papageorgiou [39, 44], by Despi et al. [33], and Biswas [28] who independently introduced the concept of an Intuitionistic Fuzzy Cognitive Map (IFCM). In [34] Peter Hadjistoykov and the author introduced the following another, more general approach to introducing of the concept of an IFCM.

Let $C = \{C_1, C_2, \cdot, C_n\}$ be a set of cognitive units and for every i ($i \in \{1, 2, \cdot, n\}$), $\mu_C(C_i)$ and $\nu_C(C_i)$ are degrees of validity and non-validity of the cognitive unit C_i.

Extending Chen's formal definitions of Fuzzy Cognitive Map (FCM, see [30]), we introduce the concept of an Intuitionistic FCM (IFCM) as the pair

$$IFCM = \langle C, E \rangle,$$

where

$$C = \{\langle C_i, \mu_C(C_i), \nu_C(C_i) \rangle | C_i \in \mathcal{C}\}$$

is an IFS and

$$E = [\mathcal{C}, \mathcal{C}, \{\langle \mu_E(e_{i,j}), \nu_E(e_{i,j}) \rangle\}],$$

is an IFIM of incidence and for every $i, j \in \{1, 2, \cdot, n\}$, $\mu_E(e_{i,j})$ and $\nu_E(e_{i,j})$ are degrees of validity and non-validity of the oriented edge between neighbouring nodes $C_i, C_j \in \mathcal{C}$.

Peter Hadjistoykov and the author extended the concept of an IFCM in two directions: in [35] we introduced the concept of an IFCM with descriptors and in [36]—Temporal IFCM.

Let T be a set of real numbers that will be interpreted as time-moments.

Let $\mathcal{C} = \{C_1, C_2, \cdot, C_n\}$ be a set of cognitive units and for every i ($i \in \{1, 2, \cdot, n\}$), $\mu_C(C_i)$ and $\nu_C(C_i)$ be the degrees of validity and non-validity of the cognitive unit C_i.

Extending the IFCM, we introduce the concept of Temporal Intuitionistic FCM (TIFCM) as the pair

$$IFCM = \langle C(T), E(T) \rangle,$$

where

$$C(T) = \{\langle C_i, \mu_C(C_i, t), \nu_C(C_i, t) \rangle | C_i \in \mathcal{C} \& t \in T\}$$

is a TIFS and

$$E(T) = [\mathcal{C}, \mathcal{C}, \{\langle \mu_E(e_{i,j}, t), \nu_E(e_{i,j}, t) \rangle\}],$$

is an Intuitionistic Fuzzy Index Matrix of incidence and for every $t \in T$, $i, j \in \{1, 2, \cdot, n\}$, $\mu_E(e_{i,j}, t)$ and $\nu_E(e_{i,j}, t)$ are degrees of validity and non-validity of the oriented edge between neighbouring nodes $C_i, C_j \in \mathcal{C}$ of the temporal IF graph (see, [6]) in time-moment t.

For every fixed time-moment $t \in T$ and for every two cognitive units C_i and C_j that are connected with an edge $e_{i,j}$, we can introduce different criteria for correctness, e.g. if C_i is higher than C_j (i.e., $\langle \mu_C(C_i, t), \nu_C(C_i, t) \rangle \geq \langle \mu_C(C_j, t), \nu_C(C_j, t) \rangle$), then

1 (top-down-max-min) $\langle \mu_C(C_i, t), \nu_C(C_i, t) \rangle \vee \langle \mu_E(e_{i,j}, t), \nu_E(e_{i,j}, t) \rangle \geq \langle \mu_C(C_j, t), \nu_C(C_j, t) \rangle$;
2 (top-down-average) $\langle \mu_C(C_i, t), \nu_C(C_i, t) \rangle @ \langle \mu_E(e_{i,j}, t), \nu_E(e_{i,j}, t) \rangle \geq \langle \mu_C(C_j, t), \nu_C(C_j, t) \rangle$;
3 (top-down-min-max) $\langle \mu_C(C_i, t), \nu_C(C_i, t) \rangle \wedge \langle \mu_E(e_{i,j}, t), \nu_E(e_{i,j}, t) \rangle \geq \langle \mu_C(C_j, t), \nu_C(C_j, t) \rangle$;
4 (bottom-up-max-min) $\langle \mu_C(C_i, t), \nu_C(C_i, t) \rangle \wedge \langle \mu_E(e_{i,j}, t), \nu_E(e_{i,j}, t) \rangle \leq \langle \mu_C(C_j, t), \nu_C(C_j, t) \rangle$;
5 (bottom-up-average) $\langle \mu_C(C_i, t) \nu_C(C_i, t) \rangle @ \langle \mu_E(e_{i,j}, t), \nu_E(e_{i,j}, t) \rangle \leq \langle \mu_C(C_j, t), \nu_C(C_j, t) \rangle$;
6 (bottom-up-min-max) $\langle \mu_C(C_i, t), \nu_C(C_i, t) \rangle \vee \langle \mu_E(e_{i,j}, t), \nu_E(e_{i,j}, t) \rangle \leq \langle \mu_C(C_j, t), \nu_C(C_j, t) \rangle$,

where for pairs $\langle a, b \rangle$ and $\langle c, d \rangle$ ($a, b, c, d, a + b, c + d \in [0, 1]$),

$$\langle a, b \rangle @ \langle c, d \rangle = \langle \frac{a+c}{2}, \frac{b+d}{2} \rangle.$$

Other criteria are also possible. For example, all above criteria can be re-written for the case, when they must be valid for any $t \in T$.

If Cr is some of the above discussed or another criterion for correctness, and if all vertices and arcs of a given TIFCM satisfy criterion Cr in a fixed time-moment $t \in T$, then this TIFCM is called (Cr, t)-correct TIFCM.

If for the same criterion Cr all vertices and arcs of the same TIFCM satisfy it in all time-moments $t \in T$, then this TIFSM is called (Cr, T)-correct TIFCM.

The validity of the following assertion is checked easily on the basis of the above definitions for correctness.

If the TIFCM is:

(a) (Cr, t)-(top-down-min-max)-correct, then it is (Cr, t)-(top-down-average)-correct and (Cr, t)-(top-down-max-min)-correct;

(b) (Cr, t)-(top-down-average)-correct, then it is (Cr, t)-(top-down-max-min)-correct;

(c) (Cr, t)-(bottom-up-max-min)-correct, then it is (Cr, t)-(bottom-up-average)-correct and (Cr, t)-(bottom-up-min-max)-correct;

(d) (Cr, t)-(bottom-up-average)-correct, then it is (Cr, t)-(bottom-up-min-max)-correct;

(e) (Cr, T)-(top-down-min-max)-correct, then it is (Cr, T)-(top-down-average)-correct and (Cr, T)-(top-down-max-min)-correct;

(f) (Cr, T)-(top-down-average)-correct, then it is (Cr, T)-(top-down-max-min)-correct;

(g) (Cr, T)-(bottom-up-max-min)-correct, then it is (Cr, T)-bottom-up-average)-correct and (Cr, T)-(bottom-up-min-max)-correct;

(h) (Cr, T)-(bottom-up-average)-correct, then it is a (Cr, T)-(bottom-up-min-max)-correct.

If the TIFCM is a (Cr, T)-correct TIFCM for a set of time-moments T, then it is a (Cr, t)-correct TIFCM for each time-moment $t \in T$ and vice versa, if it is a (Cr, t)-correct TIFCM for each separate time-moment $t \in T$, then it is a (Cr, T)-correct TIFCM.

Chapter 5
Index Matrices with Function-Type of Elements

Let the set of all used functions be $\mathcal{F}$.

The research over IMs with function-type of elements has two cases:

- each function of set $\mathcal{F}$ has one argument and it is exactly x (i.e., it is not possible that one of the functions has argument x and another function has argument y)—let us mark the set of these functions by $\mathcal{F}_x^1$;
- each function of set $\mathcal{F}$ has one argument, but that argument might be different for the different functions or the different functions of set $\mathcal{F}$ have different numbers of arguments.

Here, we discuss the two cases, simultaneously.

5.1 Definition of the Index Matrix with Function-Type of Elements

The IM with Function-type of Elements (IMFE) has the form (see [15])

$$[K, L, \{f_{k_i, l_j}\}] \equiv \begin{array}{c|ccccc} & l_1 & \cdots & l_j & \cdots & l_n \\ \hline k_1 & f_{k_1, l_1} & \cdots & f_{k_1, l_j} & \cdots & f_{k_1, l_n} \\ \vdots & \vdots & \cdots & \vdots & \cdots & \vdots \\ k_i & f_{k_i, l_1} & \cdots & f_{k_i, l_j} & \cdots & f_{k_i, l_n} \\ \vdots & \vdots & \cdots & \vdots & \cdots & \vdots \\ k_m & f_{k_m, l_1} & \cdots & f_{k_m, l_j} & \cdots & f_{k_m, l_n} \end{array},$$

where $K = \{k_1, k_2, \ldots, k_m\}$, $L = \{l_1, l_2, \ldots, l_n\}$, for $1 \le i \le m$, and $1 \le j \le n$: $f_{k_i, l_j} \in \mathcal{F}_x^1$.

The IMFE has this form independently of the form of its elements. They can be functions from $\mathcal{F}_x^1$ having one, exactly determined argument (e.g., x), as well as functions with a lot of arguments. The set of n-argument functions will be marked by $\mathcal{F}^n$.

© Springer International Publishing Switzerland 2014

K.T. Atanassov, *Index Matrices: Towards an Augmented Matrix Calculus*, Studies in Computational Intelligence 573, DOI 10.1007/978-3-319-10945-9_5

5.2 Standard Operations Over IMFEs

The forms of these operations also dependent on the forms of IMFE-elements. The definitions of four of these operations coincide with the operations over IM from Sect. 1.2, respectively. In two of the definitions, there are small differences and by this reason we give them separately.

Let the IMFEs $A = [K, L, \{f_{k_i, l_j}\}]$, $B = [P, Q, \{g_{p_r, q_s}\}]$ be given. Then

Addition:

$$A \oplus_{(\circ)} B = [K \cup P, L \cup Q, \{h_{t_u, v_w}\}],$$

where

$$
h_{t_u, v_w} =
\begin{cases}
f_{k_i, l_j}, & \text{if } t_u = k_i \in K \text{ and } v_w = l_j \in L - Q \\
& \text{or } t_u = k_i \in K - P \text{ and } v_w = l_j \in L; \\
g_{p_r, q_s}, & \text{if } t_u = p_r \in P \text{ and } v_w = q_s \in Q - L \\
& \text{or } t_u = p_r \in P - K \text{ and } v_w = q_s \in Q;, \\
f_{k_i, l_j} \circ g_{p_r, q_s}, & \text{if } t_u = k_i = p_r \in K \cap P \\
& \text{and } v_w = l_j = q_s \in L \cap Q; \\
\bot, & \text{otherwise}
\end{cases}
$$

where here and below, symbol "$\bot$"denotes the lack of operation in the respective place and '$\circ \in \{+, \times, \max, \min, \ldots\}$.

Termwise multiplication

$$A \otimes_{(\circ)} B = [K \cap P, L \cap Q, \{h_{t_u, v_w}\}],$$

where

$$h_{t_u, v_w} = f_{k_i, l_j} \circ g_{p_r, q_s},$$

for $t_u = k_i = p_r \in K \cap P$ and $v_w = l_j = q_s \in L \cap Q$.

Multiplication

$$A \odot_{(\circ, *)} B = [K \cup (P - L), Q \cup (L - P), \{c_{t_u, v_w}\}],$$

where

$$
h_{t_u, v_w} =
\begin{cases}
f_{k_i, l_j}, & \text{if } t_u = k_i \in K \\
& \text{and } v_w = l_j \in L - P - Q \\
g_{p_r, q_s}, & \text{if } t_u = p_r \in P - L - K \\
& \text{and } v_w = q_s \in Q \\
\underset{l_j = p_r \in L \cap P}{\circ} (f_{k_i, l_j} * g_{p_r, q_s}), & \text{if } t_u = k_i \in K \text{ and } v_w = q_s \in Q \\
\bot, & \text{otherwise}
\end{cases}
$$

where $(\circ, *) \in \{(+, \times), (\max, \min), (\min, \max), \ldots\}$.

Here we give an example. Let us have the IMs X and Y

$$X = \begin{array}{c|ccc} & c & d & e \\ \hline a & f_1 & f_2 & f_3 \\ b & f_4 & f_5 & f_6 \end{array}, \quad Y = \begin{array}{c|cc} & c & r \\ \hline a & g_1 & g_2 \\ p & g_3 & g_4 \\ q & g_5 & g_6 \end{array},$$

with elements

$$f_i(x) = x^i, \quad g_i(x) = \frac{1}{i.x}$$

for $i = 1, 2, \ldots, 6$. Therefore, the elements of both IMFE are elements of set $\mathcal{F}_x^1$. Hence,

$$X \oplus_{(+)} Y = \begin{array}{c|cccc} & c & d & e & r \\ \hline a & f_1 + g_1 & f_2 & f_3 & g_2 \\ b & f_4 & f_5 & f_6 & \perp \\ p & g_3 & \perp & \perp & g_4 \\ q & g_5 & \perp & \perp & g_6 \end{array} = \begin{array}{c|cccc} & c & d & e & r \\ \hline a & x + \frac{1}{x} & x^2 & x^3 & \frac{1}{2x} \\ b & x^4 & x^5 & x^6 & \perp \\ p & \frac{1}{3x} & \perp & \perp & \frac{1}{4x} \\ q & \frac{1}{5x} & \perp & \perp & \frac{1}{6x} \end{array}$$

is an IMFE with elements of set $\mathcal{F}_x^1$.

On the other hand, if for $i = 1, 2, \ldots, 6$: $g_i(y) = \frac{1}{i.y}$, then $g_i \in \mathcal{F}_y^1$ and for the above IMFE X and for the new IMFE Y we obtain:

$$X \oplus_{(+)} Y = \begin{array}{c|cccc} & c & d & e & r \\ \hline a & f_1 + g_1 & f_2 & f_3 & g_2 \\ b & f_4 & f_5 & f_6 & \perp \\ p & g_3 & \perp & \perp & g_4 \\ q & g_5 & \perp & \perp & g_6 \end{array} = \begin{array}{c|cccc} & c & d & e & r \\ \hline a & x + \frac{1}{y} & x^2 & x^3 & \frac{1}{2y} \\ b & x^4 & x^5 & x^6 & \perp \\ p & \frac{1}{3y} & \perp & \perp & \frac{1}{4y} \\ q & \frac{1}{5y} & \perp & \perp & \frac{1}{6y} \end{array},$$

i.e., the IMFE $X \oplus_{(+)} Y \in \mathcal{F}^2$. Obviously, if $g_i \in \mathcal{F}_{y,z}^1$, i.e., if it has two arguments (y and z, different from x), then $X \oplus_{(+)} Y \in \mathcal{F}_{x,y,z}^3$. It is suitable to define for each function f with n arguments: $v(f) = n$.

5.3 Relations Over IMFEs

Let everywhere, variable x obtain values in set $\mathcal{X}$ (e.g., $\mathcal{X}$ being a set of real numbers) and let $a \in \mathcal{X}$ be an arbitrary value of x.

Let the two IMFEs $A = [K, L, \{f_{k,l}\}]$ and $B = [P, Q, \{g_{p,q}\}]$ be given. We introduce the following definitions where $\subset$ and $\subseteq$ denote the relations "*strong inclusion*" and "*weak inclusion*".

The strict relation "inclusion about dimension", when the IMFE-elements of both matrices are elements of $\mathcal{F}_x^1$, is

$$A \subset_d B \text{ iff } (((K \subset P) \& (L \subset Q)) \vee ((K \subseteq P) \& (L \subset Q))$$

$$\vee ((K \subset P) \& (L \subseteq Q))) \& (\forall k \in K)(\forall l \in L)(\forall a \in X)(f_{k,l}(a) = g_{k,l}(a)).$$

The strict relation "inclusion about dimension", when the IMFE-elements of both matrices are elements not only of $\mathcal{F}_x^1$, is

$$A \subset_d B \text{ iff } (((K \subset P) \& (L \subset Q)) \vee ((K \subseteq P) \& (L \subset Q))$$

$$\vee ((K \subset P) \& (L \subseteq Q))) \& (\forall k \in K)(\forall l \in L)(v(f_{k,l}) = v(g_{k,l}))$$

$$\& (\forall a_1, \ldots, a_{v(f_{k,l})} \in X)(f_{k,l}(a_1, \ldots, a_{v(f_{k,l})}) = g_{k,l}(a_1, \ldots, a_{v(f_{k,l})})).$$

The non-strict relation "inclusion about dimension", when the IMFE-elements of both matrices are elements of $\mathcal{F}_x^1$, is

$$A \subseteq_d B \text{ iff } (K \subseteq P) \& (L \subseteq Q) \& (\forall k \in K)(\forall l \in L)(\forall a \in X)$$

$$(f_{k,l}(a) = g_{k,l}(a)).$$

The non-strict relation "inclusion about dimension", when the IMFE-elements of both matrices are elements not only of $\mathcal{F}_x^1$, is

$$A \subseteq_d B \text{ iff } (K \subseteq P) \& (L \subseteq Q) \& (\forall k \in K)(\forall l \in L)(v(f_{k,l}) = v(g_{k,l}))$$

$$\& (\forall a_1, \ldots, a_{v(f_{k,l})} \in X)(f_{k,l}(a_1, \ldots, a_{v(f_{k,l})}) = g_{k,l}(a_1, \ldots, a_{v(f_{k,l})})).$$

The strict relation "inclusion about value", when the IMFE-elements of both matrices are elements of $\mathcal{F}_x^1$, is

$$A \subset_v B \text{ iff } (K = P) \& (L = Q) \& (\forall k \in K)(\forall l \in L)(\forall a \in X)$$

$$(f_{k,l}(a) < g_{k,l}(a)).$$

The strict relation "inclusion about value", when the IMFE-elements of both matrices are elements not only of $\mathcal{F}_x^1$, is

$$A \subset_v B \text{ iff } (K = P) \& (L = Q) \& (\forall k \in K)(\forall l \in L)(v(f_{k,l}) = v(g_{k,l}))$$

$$\& (\forall a_1, \ldots, a_{v(f_{k,l})} \in X)(f_{k,l}(a_1, \ldots, a_{v(f_{k,l})}) < g_{k,l}(a_1, \ldots, a_{v(f_{k,l})})).$$

The non-strict relation "inclusion about value", when the IMFE-elements of both matrices are elements of $\mathcal{F}_x^1$, is

$$A \subseteq_v B \text{ iff } (K = P) \,\&\, (L = Q) \,\&\, (\forall k \in K)(\forall l \in L)(\forall a \in \mathcal{X})$$

$$(f_{k,l}(a) \leq g_{k,l}(a)).$$

The non-strict relation "inclusion about value", when the IMFE-elements of both matrices are elements not only of $\mathcal{F}_x^1$, is

$$A \subseteq_v B \text{ iff } (K = P) \,\&\, (L = Q) \,\&\, (\forall k \in K)(\forall l \in L)(v(f_{k,l}) = v(g_{k,l}))$$

$$\&\, (\forall a_1, \ldots, a_{v(f_{k,l})} \in \mathcal{X})(f_{k,l}(a_1, \ldots, a_{v(f_{k,l})}) \leq g_{k,l}(a_1, \ldots, a_{v(f_{k,l})})).$$

The strict relation "inclusion", when the IMFE-elements of both matrices are elements of $\mathcal{F}_x^1$, is

$$A \subset B \text{ iff } (((K \subset P) \,\&\, (L \subset Q)) \lor ((K \subseteq P) \,\&\, (L \subset Q))$$

$$\lor ((K \subset P) \,\&\, (L \subseteq Q))) \,\&\, (\forall k \in K)(\forall l \in L)(\forall a \in \mathcal{X})(f_{k,l}(a) < g_{k,l}(a)).$$

The strict relation "inclusion", when the IMFE-elements of both matrices are elements not only of $\mathcal{F}_x^1$, is

$$A \subset B \text{ iff } (((K \subset P) \,\&\, (L \subset Q)) \lor ((K \subseteq P) \,\&\, (L \subset Q))$$

$$\lor ((K \subset P) \,\&\, (L \subseteq Q))) \,\&\, (\forall k \in K)(\forall l \in L)(v(f_{k,l}) = v(g_{k,l}))$$

$$\&\, (\forall a_1, \ldots, a_{v(f_{k,l})} \in \mathcal{X})(f_{k,l}(a_1, \ldots, a_{v(f_{k,l})}) < g_{k,l}(a_1, \ldots, a_{v(f_{k,l})})).$$

The non-strict relation "inclusion", when the IMFE-elements of both matrices are elements not only of $\mathcal{F}_x^1$, is

$$A \subseteq B \text{ iff } (K \subseteq P) \,\&\, (L \subseteq Q) \,\&\, (\forall k \in K)(\forall l \in L)(\forall a \in \mathcal{X})$$

$$(f_{k,l}(a) \leq g_{k,l}(a)).$$

The non-strict relation "inclusion", when the IMFE-elements of both matrices are elements not only of $\mathcal{F}_x^1$, is

$$A \subseteq B \text{ iff } (K \subseteq P) \,\&\, (L \subseteq Q) \,\&\, (\forall k \in K)(\forall l \in L)(v(f_{k,l}) = v(g_{k,l}))$$

$$\&\, (\forall a_1, \ldots, a_{v(f_{k,l})} \in \mathcal{X})(f_{k,l}(a_1, \ldots, a_{v(f_{k,l})}) < g_{k,l}(a_1, \ldots, a_{v(f_{k,l})})).$$

All operators from Sects. 1.6–1.8 are valid without changes.

5.4 Operations Over IMFEs and IMs

Let the IM $A = [K, L, \{a_{k_i, l_j}\}]$, where $a_{k_i, l_j} \in \mathcal{R}$ and IMFE $F = [P, Q, \{f_{p_r, q_s}\}]$ be given. Then, we can define:

(a) $A \boxplus F = [K \cup P, L \cup Q, \{h_{t_u, v_w}\}]$, where

$$
h_{t_u, v_w} = \begin{cases} a_{k_i, l_j} \cdot f_{p_r, q_s}, & \text{if } t_u = k_i = p_r \in K \cap P \\ & \text{and } v_w = l_j = q_s \in L \cap Q; , \\ \perp, & \text{otherwise} \end{cases}
$$

with elements of $\mathcal{F}^1$;

(b) $A \boxtimes F = [K \cap P, L \cap Q, \{h_{t_u, v_w}\}]$, where

$$
h_{t_u, v_w} = a_{k_i, l_j} \cdot f_{p_r, q_s},
$$

for $t_u = k_i = p_r \in K \cap P$ and $v_w = l_j = q_s \in L \cap Q$ with elements of $\mathcal{F}^1$;

(c) $F \oplus A = [K \cup P, L \cup Q, \{h_{t_u, v_w}\}]$, where

$$
h_{t_u, v_w} = \begin{cases} f_{p_r, q_s}(a_{k_i, l_j}), & \text{if } t_u = k_i = p_r \in K \cap P \\ & \text{and } v_w = l_j = q_s \in L \cap Q \\ \perp, & \text{otherwise} \end{cases}
$$

with elements of $\mathcal{R}$;

(d) $F \otimes A = [K \cap P, L \cap Q, \{h_{t_u, v_w}\}]$, where

$$
h_{t_u, v_w} = f_{p_r, q_s}(a_{k_i, l_j}),
$$

for $t_u = k_i = p_r \in K \cap P$ and $v_w = l_j = q_s \in L \cap Q$ with elements of $\mathcal{R}$.

Let the IM $A = [K, L, \{\langle a_{k_i, l_j}^1, \ldots, a_{k_i, l_j}^n \rangle\}]$, for the natural number $n \geq 2$, where $a_{k_i, l_j}^1, \ldots, a_{k_i, l_j}^n \in \mathcal{R}$ and IMFE $F = [P, Q, \{f_{p_r, q_s}\}]$, where $f_{p_r, q_s} : \mathcal{F}^n \to \mathcal{F}$ be given. Then

(e) $F \diamondsuit_\oplus A = [K \cup P, L \cup Q, \{h_{t_u, v_w}\}]$, where

$$
h_{t_u, v_w} = \begin{cases} f_{p_r, q_s}(\langle a_{k_i, l_j}^1, \ldots, a_{k_i, l_j}^n \rangle), & \text{if } t_u = k_i = p_r \in K \cap P \\ & \text{and } v_w = l_j = q_s \in L \cap Q \\ \perp, & \text{otherwise} \end{cases}
$$

with elements of $\mathcal{R}$;

(f) $F \lozenge_\otimes A = [K \cap P, L \cap Q, \{h_{t_u,v_w}\}]$, where

$$h_{t_u,v_w} = f_{p_r,q_s}(\langle a^1_{k_i,l_j}, \ldots, a^n_{k_i,l_j}\rangle),$$

for $t_u = k_i = p_r \in K \cap P$ and $v_w = l_j = q_s \in L \cap Q$ with elements of $\mathcal{R}$.

The so defined IMFEs generate some new ideas. For example, when we have a sequence of real (complex) functions $f_1, f_2, \ldots, f_n \in \mathcal{F}^1_x$ and sequence of real (complex) numbers $a_1, a_2, \ldots, a_n$, then we can calculate sequentially the values $f_1(a_1), f_2(a_2), \ldots f_n(a_n)$. But, as we saw above, when these functions and these numbers are elements of IMs, we can calculate them in parallel.

The **open questions** are:

1. Will we obtain some new (additional) possibilities, if we use IMFEs?
2. To develop a theory of IMFEs with respect to the theory of functions and functional analysis.

5.5 An Example: IM-Interpretation of a Multilayer Perceptron

The artificial neural networks represent a mathematical model inspired by the biological neural networks. Its functions are borrowed from the functions of human brain. There is not yet an uniform opinion on the definition of neural networks, yet increasingly more specialists share the view that neural networks are a number of simple connected items, each featuring a rather limited local memory. These items are connected with connections, transferring numerical data, coded with various tools.

The classical three-layered neural network, in abbreviated notation, has the form

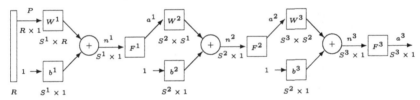

In multilayered networks, the exits of one layer become entries for the next one. The equations describing this operation are:

$$a^{m+1} = f^{m+1}(w^{m+1}.a^m + b^{m+1})$$

for $m = 0, 1, 2, \ldots, M - 1$, where:

- m is the current number of the layers in the network;
- M is the number of the layers in the network;
- P is an entry networks vector;
- a^m is the exit of the m-th layer of the neural network;

- s^m is a number of neutrons of a m-th layer of the neural network;
- W is a matrix of the coefficients of all inputs;
- b is neurons input bias;
- F^m is the transfer function of the m-th layer exit.

Now, we describe the IM-representation of the results of the work on the above multilayered network.

Let P be an input vector in the form

$$P = \frac{\begin{array}{ccc} p_1 & \cdots & p_R \end{array}}{p_0 \,\big|\, \begin{array}{ccc} s_1 & \cdots & s_R \end{array}}.$$

Let the weight coefficients of the connections between the nodes of the input vector and these from the first layer be given by the IM

$$W^1 = \begin{array}{c|ccc} & a_{1,1} & \cdots & a_{1,s_1} \\ \hline p_1 & W^1_{1,1} & \cdots & W^1_{1,s_1} \\ \vdots & \vdots & \cdots & \vdots \\ p_R & W^1_{R,1} & \cdots & W^1_{R,s_1} \end{array},$$

while let the parameters of the moves of the neurons from the first layer be given by the IM

$$B^1 = \frac{\begin{array}{ccc} a_{1,1} & \cdots & a_{1,s_1} \end{array}}{p_0 \,\big|\, \begin{array}{ccc} b_{1,1} & \cdots & b_{1,s_1} \end{array}}.$$

Then, a_1 is the IM with the values of the neurons in the first layer. It is obtained by the formula

$$a^1 = (P \odot W^1) \oplus B^1$$

$$= \frac{\begin{array}{ccc} a_{1,1} & \cdots & a_{1,s_1} \end{array}}{p_0 \,\big|\, \begin{array}{ccc} \displaystyle\sum_{k=1}^{R}(a_k W^1_{k,1} + b_{k,1}) & \cdots & \displaystyle\sum_{k=1}^{R}(a_k W^1_{k,s_1} + b_{k,s_1}) \end{array}}$$

$$= \frac{\begin{array}{ccc} a_{1,1} & \cdots & a_{1,s_1} \end{array}}{p_0 \,\big|\, \begin{array}{ccc} a^1_1 & \cdots & a^1_{s_1} \end{array}}.$$

Let i be a natural number from the set $\{2, 3, \ldots, M\}$. Let the IM of the weight coefficients of the connections between the nodes of the i-th and $(i+1)$-st layers be

$$W^i = \begin{array}{c|ccc} & a_{1,1} & \cdots & a_{1,s_i} \\ \hline a_{i-1,1} & W^{i-1}_{1,1} & \cdots & W^{i-1}_{1,s_i} \\ \vdots & \vdots & \cdots & \vdots \\ a_{i-1,s_i-1} & W^{i-1}_{s_i-1,1} & \cdots & W^{i-1}_{s_i-1,s_i} \end{array}$$

and let the parameters of the moves of the neurons from the i-th layer be given by the IM

$$B^i = \frac{\begin{array}{ccc} a_{1,i} & \cdots & a_{1,s_i} \end{array}}{p_0 \; \begin{array}{ccc} b_{i,1} & \cdots & b_{i,s_i} \end{array}}.$$

Let us have the IM for the $(i\,1)$-st layer

$$a^{i-1} = \frac{\begin{array}{ccc} a_{1,1} & \cdots & a_{1,s_i-1} \end{array}}{p_0 \; \begin{array}{ccc} a_1^{i-1} & \cdots & a_{s_i-1}^{i-1} \end{array}}.$$

Then

$$a^i = (a^{i-1} \odot W^i) \oplus B^i$$

$$= \frac{\begin{array}{ccc} a_{1,1} & \cdots & a_{1,s_i} \end{array}}{p_0 \; \begin{array}{ccc} \sum\limits_{k=1}^{R}(a_k^{i-1}W_{k,1}^i + b_{k,1}^i) & \cdots & \sum\limits_{k=1}^{R}(a_k^{i-1}W_{k,s_i}^i + b_{k,s_i}^i) \end{array}}$$

$$= \frac{\begin{array}{ccc} a_{1,1} & \cdots & a_{1,s_i} \end{array}}{p_0 \; \begin{array}{ccc} a_1^i & \cdots & a_{s_i}^i \end{array}}.$$

and for $i = M$

$$a^{M-1} = \frac{\begin{array}{ccc} a_{M,1} & \cdots & a_{M,s_{M-1}} \end{array}}{p_0 \; \begin{array}{ccc} a_1^{M-1} & \cdots & a_{s_{M-1}}^{M-1} \end{array}}.$$

In [38], the transfer function $F : \mathcal{R} \to \mathcal{R}$ is defined. We see that the above formulas can be interpreted as results of identical function F, i.e., for every real number x, $F(x) = x$. Below, firstly, we discuss the case, when F is a real function, different from the identical one. For example, in [37] this function is sigmoidal or hyperbolical tangent function.

Now, following [23], for a fixed real function F, we define over the IM

$$A = \frac{\begin{array}{cccc} l_1 & l_2 & \cdots & l_n \end{array}}{\begin{array}{c} k_1 \\ k_2 \\ \vdots \\ k_m \end{array} \begin{array}{cccc} a_{k_1,l_1} & a_{k_1,l_2} & \cdots & a_{k_1,l_n} \\ a_{k_2,l_1} & a_{k_2,l_2} & \cdots & a_{k_2,l_n} \\ & & & \\ a_{k_m,l_1} & a_{k_m,l_2} & \cdots & a_{k_m,l_n} \end{array}}$$

the operator

$$O_F(A) = \begin{array}{c|cccc} & l_1 & l_2 & \dots & l_n \\ \hline k_1 & F(a_{k_1,l_1}) & F(a_{k_1,l_2}) & \dots & F(a_{k_1,l_n}) \\ k_2 & F(a_{k_2,l_1}) & F(a_{k_2,l_2}) & \dots & F(a_{k_2,l_n}) \\ \vdots & & & & \\ k_m & F(a_{k_m,l_1}) & F(a_{k_m,l_2}) & \dots & F(a_{k_m,l_n}) \end{array}.$$

Hence, we can describe the neural network with the form

$$a^1 = O_F((P \odot W^1) \oplus B^1),$$

$$a^i = O_F((a^{i-1} \odot W^i) \oplus B^i).$$

Therefore,

$$a^{M-1} = O_F((a^{M-2} \odot W^{M-1}) \oplus B^{M-1})$$

$$= O_F((\dots O_F((O_F((P \odot W^1) \oplus B^1) \oplus B^2)\dots \oplus B^{M-2}) \oplus B^{M-1}).$$

A more general case is the following: each layer hat its own transfer function, i.e., function F_i is associated to the i-th layer. Therefore, the NN has the IM-representation

$$a^{M-1} = O_{F_{M-1}}((\dots O_{F_2}((O_{F_1}((P \odot W^1) \oplus B^1) \oplus B^2)\dots \oplus B^{M-2}) \oplus B^{M-1}).$$

Below, we will extend the results from [23], using the ideas from Sect. 5.4. Now, for each layer we juxtapose an IMFE

$$F^i = \begin{array}{c|ccc} & a_{1,1} & \dots & a_{1,s_i} \\ \hline p_0 & f_{1,1} & \dots & f_{1,s_i} \end{array},$$

where $f_{i,j} \in \mathcal{F}_x^1$ for $1 \le i \le M-1$ and $1 \le j \le s_i$. Therefore, for the j-th node from i-th layer of the multilayered network we juxtapose the function $f_{i,j}$ and in a result, we obtain

$$a^1 = F_1 \oplus ((P \odot W^1) \oplus B^1),$$

$$a^i = F_i \oplus ((a^{i-1} \odot W^i) \oplus B^i).$$

Therefore,

$$a^{M-1} = F_{M-1} \oplus ((a^{M-2} \odot W^{M-1}) \oplus B^{M-1})$$

$$= F_{M-1} \oplus ((\dots F_2 \oplus ((F_1 \oplus ((P \odot W^1) \oplus B^1) \oplus B^2)\dots \oplus B^{M-2}) \oplus B^{M-1}).$$

We will finish with the following **Open problems**

7. What other IM-interpretations of the standard and modified NNs are appropriate?
8. Can IM-interpretations of genetic algorithms procedures be designed?
9. Can IM-interpretations of ant colony optimization procedures be designed?
10. Can IM-interpretations of game method for modelling (see [21]) procedures be designed?

Chapter 6
Three Dimensional IMs

6.1 Definition of a Three Dimensional IMs

The TIFIMs, ETIFIMs and IMFEs show the necessity for definition of the concept of a three dimensional IM (3D-IM).

Let $\mathcal{I}$ be a fixed set of indices and $\mathcal{X}$ be a fixed set of objects. Following [21], we call "3D-IM"with index sets K, L and H (K, L, $H \subset \mathcal{I}$) the object:

$$[K, L, H, \{a_{k_i, l_j, h_g}\}]$$

$$\equiv \left\{ \begin{array}{c|cccccc} h_g & l_1 & \ldots & l_j & \ldots & l_n \\ \hline k_1 & a_{k_1,l_1,h_g} & \vdots & a_{k_1,l_j,h_g} & \cdots & a_{k_1,l_n,h_g} \\ \vdots & \vdots & \ldots & \vdots & \ldots & \vdots \\ k_m & a_{k_m,l_1,h_g} & \cdots & a_{k_m,l_j,h_g} & \cdots & a_{k_m,l_n,h_g} \end{array} \middle| h_g \in H \right\}$$

$$\equiv \left\{ \begin{array}{c|cccccc} h_1 & l_1 & \ldots & l_j & \ldots & l_n \\ \hline k_1 & a_{k_1,l_1,h_1} & \vdots & a_{k_1,l_j,h_1} & \cdots & a_{k_1,l_n,h_1} \\ \vdots & \vdots & \ldots & \vdots & \ldots & \vdots \\ k_m & a_{k_m,l_1,h_1} & \cdots & a_{k_m,l_j,h_1} & \cdots & a_{k_m,l_n,h_1} \end{array} \right. ,$$

$$\begin{array}{c|cccccc} h_2 & l_1 & \ldots & l_j & \ldots & l_n \\ \hline k_1 & a_{k_1,l_1,h_2} & \vdots & a_{k_1,l_j,h_2} & \cdots & a_{k_1,l_n,h_2} \\ \vdots & \vdots & \ldots & \vdots & \ldots & \vdots \\ k_m & a_{k_m,l_1,h_2} & \cdots & a_{k_m,l_j,h_2} & \cdots & a_{k_m,l_n,h_2} \end{array} ,$$

© Springer International Publishing Switzerland 2014
K.T. Atanassov, *Index Matrices: Towards an Augmented Matrix Calculus*,
Studies in Computational Intelligence 573, DOI 10.1007/978-3-319-10945-9_6

$$
\cdots, \quad
\left.
\begin{array}{c|cccccc}
h_f & l_1 & \cdots & l_j & \cdots & l_n \\
\hline
k_1 & a_{k_1,l_1,h_f} & \vdots & a_{k_1,l_j,h_f} & \cdots & a_{k_1,l_n,h_f} \\
\vdots & \vdots & \cdots & \vdots & \cdots & \vdots \\
k_m & a_{k_m,l_1,h_f} & \cdots & a_{k_m,l_j,h_f} & \cdots & a_{k_m,l_n,h_f}
\end{array}
\right],
$$

where $K = \{k_1, k_2, \ldots, k_m\}$, $L = \{l_1, l_2, \ldots, l_n\}$, $H = \{h_1, h_2, \ldots, h_f\}$, and for $1 \leq i \leq m, 1 \leq j \leq n, 1 \leq g \leq f : a_{k_i,l_j,h_g} \in \mathcal{X}$.

6.2 Operations Over 3D-IMs

First, we start with operation "transposition".

As we saw in Sect. 3.7, there are $2 (= 2!)$ EIM, related to this operation: the standard EIM and its transposed EIM. Now, for 3D-IMs, there are $6 (=3!)$ cases: the standard 3D-IM and five different transposed 3D-IMs. The geometrical and analytical forms of the separate transposed 3D-IMs are the following.

[1,2,3]-transposition (identity)

$$
[K, L, H, \{a_{k_i,l_j,h_g}\}]^{[1,2,3]} = [K, L, H, \{a_{k_i,l_j,h_g}\}];
$$

[1,3,2]-transposition

$$
[K, L, H, \{a_{k_i,l_j,h_g}\}]^{[1,3,2]} = [K, H, L, \{a_{k_i,h_g,l_j}\}];
$$

[2,1,3]-transposition

$$[K, L, H, \{a_{k_i,l_j,h_g}\}]^{[2,1,3]} = [L, K, H, \{a_{l_j,k_i,h_g}\}];$$

[2,3,1]-transposition

$$[K, L, H, \{a_{k_i,l_j,h_g}\}]^{[2,3,1]} = [L, H, K, \{a_{l_j,h_g,k_i}\}];$$

[3,1,2]-transposition

$$[K, L, H, \{a_{k_i,l_j,h_g}\}]^{[3,1,2]} = [H, K, L, \{a_{h_g,k_i,l_j}\}];$$

[3,2,1]-**transposition**

$$\left(\begin{array}{c} \begin{array}{cc} H \diagup & \\ K \begin{array}{|c} L \end{array} & \end{array} \end{array} \right) \overset{[3,2,1]}{=} \begin{array}{cc} K \diagup & \\ H \begin{array}{|c} L \end{array} & \end{array}$$

$$[K, L, H, \{a_{k_i,l_j,h_g}\}]^{[3,2,1]} = [H, L, K, \{a_{h_g,l_j,k_i}\}];$$

For the 3D-IMs $A = [K, L, H, \{a_{k_i,l_j,h_g}\}]$, $B = [P, Q, R, \{b_{p_r,q_s,e_d}\}]$, operations that are analogous of the usual matrix operations of addition and multiplication are defined, as well as other, specific ones.

Addition

$$A \oplus_{(\circ)} B = [K \cup P, L \cup Q, H \cup R, \{c_{t_u,v_w,x_y}\}],$$

where

$$c_{t_u,v_w,x_y}$$

$$= \begin{cases} a_{k_i,l_j,h_g}, & \text{if } t_u = k_i \in K, v_w = l_j \in L \text{ and } x_y = h_g \in H - R \\ & \text{or } t_u = k_i \in K, v_w = l_j \in L - Q \text{ and } x_y = h_g \in H \\ & \text{or } t_u = k_i \in K - P, v_w = l_j \in L \text{ and } x_y = h_g \in H \\[2mm] b_{p_r,q_s,e_d}, & \text{if } t_u = p_r \in P, v_w = q_s \in Q \text{ and } x_y = e_d \in R - H \\ & \text{or } t_u = p_r \in P, v_w = q_s \in Q - L \text{ and } x_y = e_d \in R \\ & \text{or } t_u = p_r \in P - K, v_w = q_s \in Q \text{ and } x_y = e_d \in R \\[2mm] a_{k_i,l_j,h_g} \circ b_{p_r,q_s,e_d}, & \text{if } t_u = k_i = p_r \in K \cap P, v_w = l_j = q_s \in L \cap Q \\ & \text{and } x_y = h_g = e_d \in H \cap R \\[2mm] 0, & \text{otherwise} \end{cases}$$

Termwise multiplication

$$A \otimes_{(\circ)} B = [K \cap P, L \cap Q, H \cap R, \{c_{t_u,v_w,x_y}\}],$$

where

$$c_{t_u,v_w,x_y} = a_{k_i,l_j,h_g} \circ b_{p_r,q_s,e_d},$$

for $t_u = k_i = p_r \in K \cap P$, $v_w = l_j = q_s \in L \cap Q$ and $x_y = h_g = e_d \in H \cap R$;

Multiplication

This operation is related to the operation "transposition". There are six different operations "multiplication". The first (standard) multiplication is

$$A \odot_{(o,*)} B = A \odot_{(o,*)}^{[1,2,3]} B = [K \cup (P - L),\, Q \cup (L - P),\, H \cup R,\, \{c_{t_u,v_w,x_y}\}],$$

where

$$c_{t_u,v_w,x_y}$$

$$=
\begin{cases}
a_{k_i,l_j,h_g}, & \text{if } t_u = k_i \in K \\
& \quad \& v_w = l_j \in L - P - Q \& x_y = h_g \in H \\
& \text{or } t_u = k_i \in K - P - Q \\
& \quad \& v_w = l_j \in L \& x_y = h_g \in H \\[2ex]
b_{pr,qs,ed}, & \text{if } t_u = p_r \in P \\
& \quad \& v_w = q_s \in Q \& x_y = e_d \in R \\
& \text{or } t_u = p_r \in P - L - K \\
& \quad \& v_w = q_s \in Q \& x_y = e_d \in R \\[2ex]
\underset{l_j=p_r \in L \cap P}{\overset{\circ}{}} a_{k_i,l_j,h_g} * b_{pr,qs,ed}, & \text{if } t_u = k_i \in K \& v_w = q_s \in Q \\
& \quad \& x_y = h_g = e_d \in H \cap R \\[2ex]
\perp, & \text{otherwise}
\end{cases}$$

The geometrical interpretation of this operation is

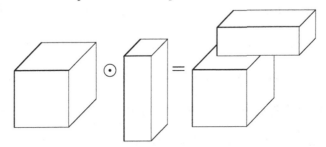

Let $[x, y, z]$ be a permutation of triple $[1, 2, 3]$. Operation $[x, y, z]$-multiplication is defined by:

$$A \odot_{(o,*)}^{[x,y,z]} B = A \odot_{(o,*)} B^{[x,y,z]}.$$

Structural subtraction

$$A \ominus B = [K - P,\, L - Q,\, H - R,\, \{c_{t_u,v_w,x_y}\}],$$

where "−" is the set-theoretic difference operation and

$$c_{t_u,v_w,x_y} = a_{k_i,l_j,h_g}, \text{ for } t_u = k_i \in K - P, v_w = l_j \in L - Q \text{ and } x_y = h_g \in H - R.$$

Multiplication with a constant

$$\alpha \cdot A = [K, L, H, \{\alpha \cdot a_{k_i,l_j,h_g}\}],$$

where α is a constant.

Termwise subtraction

$$A -_{(\circ)} B = A \oplus_{(\circ)} (-1) \cdot B.$$

The "zero"-3D-IM is

$$I_\emptyset = [\emptyset, \emptyset, \emptyset, \{a_{k_i,l_j,x_y}\}].$$

6.3 Relations Over 3D-IMs

Let the two 3D-IMs $A = [K, L, H, \{a_{k,l,h}\}]$ and $B = [P, Q, R, \{b_{p,q,e}\}]$ be given. We will introduce the following definitions where $\subset$ and $\subseteq$ denote the relations *"strong inclusion"* and *"weak inclusion"*.

The strict relation "inclusion about dimension" is

$$A \subset_d B \text{iff}(((K \subset P)\&(L \subset Q) \& (H \subset R)) \vee ((K \subseteq P)\&(L \subset Q)\&(H \subset R))$$

$$\vee((K \subset P)\&(L \subseteq Q)\&(H \subset R)) \vee ((K \subset P)\&(L \subset Q)\&(H \subseteq R)))$$

$$\&(\forall k \in K)(\forall l \in L)(\forall h \in H)(a_{k,l,h} = b_{k,l,h}).$$

The non-strict relation "inclusion about dimension" is

$$A \subseteq_d B \text{iff}(K \subseteq P)\&(L \subseteq Q)\&(H \subseteq R)\&(\forall k \in K)(\forall l \in L)$$

$$(\forall h \in H)(a_{k,l,h} = b_{k,l,h}).$$

The strict relation "inclusion about value" is

$$A \subset_v B \text{iff}(K = P)\&(L = Q)\&(H = R)\&(\forall k \in K)(\forall l \in L)(\forall h \in H)$$

$$(a_{k,l,h} < b_{k,l,h}).$$

The non-strict relation "inclusion about value" is

$$A \subseteq_v B \text{iff}(K = P)\&(L = Q)\&(H = R)\&(\forall k \in K)(\forall l \in L)(\forall h \in H)$$

$$(a_{k,l,h} \leq b_{k,l,h}).$$

The strict relation "inclusion" is

$$A \subset B \text{iff}(((K \subset P)\&(L \subset Q)\&(H \subset R)) \vee ((K \subseteq P)\&(L \subset Q)\&(H \subset R))$$

$$\vee((K \subset P)\&(L \subseteq Q)\&(H \subset R)) \vee ((K \subset P)\&(L \subset Q)\&(H \subseteq R)))$$

$$\&(\forall k \in K)(\forall l \in L)(\forall h \in H)(a_{k,l,h} < b_{k,l,h}).$$

The non-strict relation "inclusion" is

$$A \subseteq B \text{iff}(K \subseteq P)\&(L \subseteq Q)\&(H \subseteq R)\&(\forall k \in K)(\forall l \in L)(\forall h \in H)$$

$$(a_{k,l,h} \leq b_{k,l,h}).$$

6.4 Operations "Reduction" Over an 3D-IM

First, we introduce operations $(k, \perp, \perp)$-, $(\perp, l, \perp)$- and $(\perp, \perp, h)$-reduction of a given 3D-IM $A = [K, L, H, \{a_{k_i,l_j,h_g}\}]$:

$$A_{(k,\perp,\perp)} = [K - \{k\}, L, H, \{c_{t_u,v_w,h_g}\}]$$

where

$$c_{t_u,v_w,x_y} = a_{k_i,l_j,h_g} \text{ for } t_u = k_i \in K - \{k\}, v_w = l_j \in L \text{ and } x_y = h_g \in H,$$

$$A_{(\perp,l,\perp)} = [K, L - \{l\}, H, \{c_{t_u,v_w,x_y}\}],$$

where

$$c_{t_u,v_w,x_y} = a_{k_i,l_j,h_g} \text{ for } t_u = k_i \in K, v_w = l_j \in L - \{l\} \text{ and } x_y = h_g \in H$$

and

$$A_{(\perp,\perp,h)} = [K, L, H - \{h\}, \{c_{t_u,v_w,x_y}\}],$$

where

$$c_{t_u,v_w,x_y} = a_{k_i,l_j,h_g} \text{ for } t_u = k_i \in K, v_w = l_j \in L \text{ and } x_y = h_g \in H - \{h\}.$$

Second, we define

$$A_{(k,l,h)} = ((A_{(k,\perp,\perp)})_{(\perp,l,\perp)})_{(\perp,\perp,h)},$$

i.e.,

$$A_{(k,l,h)} = [K - \{k\}, L - \{l\}, H - \{h\}, \{c_{t_u,v_w,x_y}\}],$$

where

$$c_{t_u,v_w,x_y} = a_{k_i,l_j,h_g}$$

for $t_u = k_i \in K - \{k\}$, $v_w = l_j \in L - \{l\}$ and $x_y = h_g \in H - \{h\}$.

For every 3D-IM A and for every $k_1, k_2 \in K, l_1, l_2 \in L, h_1, h_2 \in H$,

$$(A_{(k_1,l_1,h_1)})_{(k_2,l_2,h_2)} = (A_{(k_2,l_2,h_2)})_{(k_1,l_1,h_1)}.$$

Third, let $P = \{p_1, p_2, \ldots, p_s\} \subseteq K, Q = \{q_1, q_2, \ldots, q_t\} \subseteq L$ and $R = \{r_1, r_2, \ldots, r_u\} \subseteq H, p \in K, l \in L, h \in H$. Now, we define the following four operations:

$$A_{(P,l,h)} = (\ldots ((A_{(p_1,l,h)})_{(p_2,l,h)}) \ldots)_{(p_s,l,h)},$$

$$A_{(k,Q,h)} = (\ldots ((A_{(k,l_1,h)})_{(k,l_2,h)}) \ldots)_{(k,l_t,h)},$$

$$A_{(k,q,H)} = (\ldots ((A_{(k,l,r_1)})_{(k,l,r_2)}) \ldots)_{(k,l,r_u)},$$

$$A_{(P,Q,H)} = (\ldots ((A_{(p_1,Q,H)})_{(p_2,Q,H)}) \ldots)_{(p_s,Q,H)}$$

$$= (\ldots ((A_{(P,q_1,H)})_{(P,q_2,H)}) \ldots)_{(P,q_t,H)} = (\ldots ((A_{(P,Q,r_1)})_{(P,Q,r_2)}) \ldots)_{(P,Q,r_u)}.$$

Obviously,

$$A_{(K,L,H)} = I_\emptyset,$$

$$A_{(\emptyset,\emptyset,\emptyset)} = A.$$

6.5 Operation "Projection" Over an IM

Let $P \subseteq K, Q \subseteq L, R \subseteq H$. Then,

$$pr_{P,Q,R} A = [P, Q, R, \{b_{k_i,l_j,h_g}\}],$$

where

$$(\forall k_i \in P)(\forall l_j \in Q)(\forall h_g \in R)(b_{k_i,l_j,h_g} = a_{k_i,l_j,h_g}).$$

6.6 Operation "Substitution" Over an 3D-IM

Let the 3D-IM $A = [K, L, H, \{a_{k,l,h}\}]$ be given.

First, local substitution over the IM is defined for the pairs of indices (p, k) and/or (q, l) and/or (r, h), respectively, by

$$\left[\frac{p}{k}; \bot; \bot\right] A = \left[(K - \{k\}) \cup \{p\}, L, H, \{a_{k_i, l_j, h_g}\}\right],$$

$$\left[\bot; \frac{q}{l}; \bot\right] A = \left[K, (L - \{l\}) \cup \{q\}, H, \{a_{k_i, l_j, h_g}\}\right],$$

$$\left[\bot; \bot; \frac{r}{h}\right] A = \left[K, L, (H - \{h\}) \cup \{r\}, \{a_{k_i, l_j, h_g}\}\right].$$

Second,

$$\left[\frac{p}{k}; \frac{q}{l}; \frac{r}{h}\right] A = \left[\frac{p}{k}; \bot; \bot\right]\left[\bot; \frac{q}{l}; \bot\right]\left[\bot; \bot; \frac{r}{h}\right] A$$

$$= \left[(K - \{k\}) \cup \{p\}, (L - \{l\}) \cup \{q\}, (H - \{h\}) \cup \{r\}, \{a_{k_i, l_j, h_g}\}\right].$$

Let the sets of indices $P = \{p_1, p_2, \ldots, p_m\}$, $Q = \{q_1, q_2, \ldots, q_n\}$, $R = \{r_1, r_2, \ldots, r_s\}$ be given, where $m = card(K), n = card(L), s = card(H)$.

Third, for them we define sequentially:

$$\left[\frac{P}{K}; \bot; \bot\right] A = \left[\frac{p_1}{k_1} \frac{p_2}{k_2} \cdots \frac{p_m}{k_m}; \bot; \bot\right] A,$$

$$\left[\bot; \frac{Q}{L}; \bot\right] A = \left[\bot; \frac{q_1}{l_1} \frac{q_2}{l_2} \cdots \frac{q_n}{l_n}; \bot\right] A,$$

$$\left[\bot; \bot; \frac{R}{H}\right] A = \left[\bot; \frac{r_1}{h_1} \frac{r_2}{h_2} \cdots \frac{r_s}{h_s}; \bot\right] A,$$

$$\left[\frac{P}{K}; \frac{Q}{L}; \frac{R}{H}\right] A = \left[\frac{p_1}{k_1} \frac{p_2}{k_2} \cdots \frac{p_m}{k_m}; \frac{q_1}{l_1} \frac{q_2}{l_2} \cdots \frac{q_n}{l_n}; \frac{r_1}{h_1} \frac{r_2}{h_2} \cdots \frac{r_s}{h_s}\right] A = [P, Q, R, \{a_{k,l,h}\}].$$

6.7 An Example with Bookshops

Let us have bookshops $B_1, B_2, \ldots, B_b$ in different towns $C_1, C_2, \ldots, C_c$. Obviously, some bookshops can be in one company and in different towns. Let us interested in the sales of the books with titles $A_1, A_2, \ldots, A_a$.

First, we can construct an 3D-IM with elements—real (natural) numbers, e.g., with the form

$$M = [A, B, C, \{d_{k_i, l_j, h_g}\}]$$

$$\equiv \left\{ \begin{array}{c|ccccc} C_g & B_1 & \cdots & B_j & \cdots & B_n \\ \hline A_1 & d_{A_1,B_1,C_g} & \vdots & d_{A_1,B_j,C_g} & \cdots & d_{A_1,B_n,C_g} \\ \vdots & \vdots & \cdots & \vdots & \cdots & \vdots \\ A_i & d_{A_i,B_1,C_g} & \vdots & d_{A_i,B_j,C_g} & \cdots & d_{A_i,B_n,C_g} \\ \vdots & \vdots & \cdots & \vdots & \cdots & \vdots \\ A_m & d_{A_m,B_1,C_g} & \cdots & d_{A_m,B_j,C_g} & \cdots & d_{A_m,B_n,C_g} \end{array} \right. \quad \Bigg| \; C_g \in C \Bigg\}$$

where $A = \{A_1, A_2, \ldots, A_a\}$, $B = \{B_1, B_2, \ldots, B_b\}$, $C = \{C_1, C_2, \ldots, C_c\}$, and for $1 \leq i \leq a, 1 \leq j \leq b, 1 \leq g \leq c : d_{A_i,B_j,C_g} \geq 0$ is a natural number, representing the total number of sold books with title A_i in bookshop B_j in town C_g.

Second, we can modify the present IM, changing its elements d_{A_i,B_j,C_g} with the IFPs

$$\langle m_{A_i,B_j,C_g}, n_{A_i,B_j,C_g} \rangle,$$

where m_{A_i,B_j,C_g} is the quantity of sold books from A_ith title, divided by the total quantity of this book, received in bookshop B_j in town C_g and n_{A_i,B_j,C_g} is the quantity of the same book in the bookshop warehouse, divided by the total quantity of this book, as above. Therefore, $m_{A_i,B_j,C_g} + n_{A_i,B_j,C_g} \leq 1$ and number $1 - m_{A_i,B_j,C_g} - n_{A_i,B_j,C_g}$ corresponds to the number of non-sold books that stay on the shelves in the bookshop, but are not in its warehouse. Therefore, the above standard 3D-IM is transformed to 3D-IFIM.

The example is interesting, because it is a good illustration not only of the possibility to transform a standard 3D-IM to a 3D-IFIM, but on its basis we can construct a 4D-IM or 4D-IFIM. For this aim, we add a fourth component in the IM-definition, e.g. – finite time $T = \{T_1, T_2, \ldots, T_t\}$. So, we obtain two new IMs with the forms

$$M_{standart\ IM} = [A, B, C, T, \{d_{A_i,B_j,C_g,t_u}\}]$$

and

$$M_{IFIM} = [A, B, C, T, \{\langle m_{A_i,B_j,C_g,t_u}, n_{A_i,B_j,C_g,t_u} \rangle\}],$$

where for the above discussed d_{A_i,B_j,C_g} and for $1 \leq u \leq t$: d_{A_i,B_j,C_g,t_u} is a natural number and $\langle m_{A_i,B_j,C_g,t_u}, n_{A_i,B_j,C_g,t_u} \rangle$ is an IFP.

The example shows that the 3D-IMs can be used for describing of databases, data warehouses and OLAP-structures. Therefore, in future, the 3D- and nD-IMs can obtain real applications in information technologies.

We will finish with the following **Open problems**

11. To develop a complete theory of 3D-IM and nD-IM, where $n \geq 3$ is a natural number.

12. To find more applications of 3D- and nD-IMs?

The solution of each one of the formulated problems will lead to the development of IM theory.

References

1. Alexieva, J., Choy, E., Koycheva, E.: Review and bibliography on generalized nets theory and applications. In: Choy, E., Krawczak, M., Shannon, A., Szmidt, E. (eds.) A Survey of Generalized Nets , Raffles KvB Monograph, vol. 10, pp. 207–301. World Scientific, Singapore (2007)
2. Atanassov, K.: Conditions in generalized nets. Proceedings of the XIII Spring Conference of the Union of Bulgarian Mathematical, Sunny Beach, pp. 219–226. (1984)
3. Atanassov, K.: Generalized index matrices. C. R. Acad Bulgare Sci. **40**(11), 15–18 (1987)
4. Atanassov, K.: Generalized Nets. World Scientific, Singapore (1991)
5. Atanassov, K.: Index matrix representation of the intuitionistic fuzzy graphs. Proceeding of Fifth Scientific Session of the Math. Foundations of Artificial Intelligence Seminar, Sofia, pp. 36–41. (1994). Accessed 5 Oct 1994 (Preprint MRL-MFAIS-10-94)
6. Atanassov, K.: Temporal intuitionistic fuzzy graphs. Notes Intuitionistic Fuzzy Sets **4**(4), 59–61 (1998)
7. Atanassov, K.: Intuitionistic Fuzzy Sets. Springer, Heidelberg (1999)
8. Atanassov, K. On index matrix interpretations of intuitionistic fuzzy graphs. Notes Intuitionistic Fuzzy Sets, **8**(4), 73–78 (2002)
9. Atanassov, K.: On Generalized Nets Theory "Prof. M. Drinov". Academic Publ. House, Sofia (2007)
10. Atanassov, K.: On index matrices. Part 1: Standard cases. Adv. Stud. Contem. Math. **20**(2), 291–302 (2010)
11. Atanassov, K.: On index matrices, Part 2: Intuitionistic fuzzy case. In: Proceedings of the Jangjeon Mathematical Society, vol. 13, issue 2, 121–126 (2010)
12. Atanassov, K.: Game method for modelling. "Prof. M. Drinov" Academic Publ. House, Sofia (2011)
13. Atanassov, K.: On Intuitionistic Fuzzy Sets Theory. Springer, Berlin (2012)
14. Atanassov, K.: On index matrices, Part 3: On the hierarchical operation over index matrices. Adv. Stud. Contem. Math. **23**(2), 225–231 (2013)
15. Atanassov, K.: Index matrices with function-type of elements. Int. J. Inf. Models Anal. **3**(2), 103–112 (2014)
16. Atanassov, K.: On extended intuitionistic fuzzy index matrices. Notes Intuitionistic Fuzzy Sets **19**(4), 27–41 (2013)
17. Atanassov, K.: Index matrices with function-type of elements. Int. J. Inf. Models Analyses **3**(2), 103–112 (2014)
18. Atanassov, K.: Extended index matrices. Proceedings of 7-th IEEE Conference "Intelligent Systems", Warsaw, pp. 24–26 (2014) (in press)
19. Atanassov, K.: Extended intuitionistic fuzzy graphs. Notes Intuitionistic Fuzzy Sets **20**(1), 1–11 (2014)

20. Atanassov, K.: Extended temporal intuitionistic fuzzy index matrix. Notes Intuitionistic Fuzzy Sets **20**(3), (2014) (in press)
21. Atanassov, K.: On index matrices. Part 5: three dimensional index matrices. Adv. Stud. Contemp. Math. (in press)
22. Atanassov, K., Mavrov, D., Atanassova, V.: A new approach for multicriteria decision making, based on index matrices and intuitionistic fuzzy sets. Modern Approaches in Fuzzy Sets, Intuitionistic Fuzzy Sets, Generalized Nets and Related Topics. Volume I: Foundations, IBS PAN—SRI PAS, Warsaw, (2014) (in press)
23. Atanassov, K., Sotirov, S.: Index matrix interpretation of the Multilayer perceptron. Proceeding of IEEE International Symposium on Innovations in Intelligent Systems and Applications (INISTA), pp. 13710966. Albena, INSPEC, (2013) Accessed 19–21 June 2013
24. Atanassov, K., Sotirova, E., Bureva, V.: On index matrices. Part 4: new operations over index matrices. Adv. Stud. Contemp. Math. **23**(3), 547–552 (2013)
25. Atanassov, K., Sotirova, E., Bureva, V., Shannon, A.: Temporal intuitionistic fuzzy index matrices. Issues Intuitionistic Fuzzy Sets Generalized Nets **10**, 54–65 (2013)
26. Atanassov, K., Szmidt, E., Kacprzyk, J.: On intuitionistic fuzzy pairs. Notes Intuitionistic Fuzzy Sets **19**(3) 1–13 (2013)
27. Axelrod, R.: Structure of Decision: The Cognitive Maps of Political Elites. Princeton Univ. Press, New Jersey (1976)
28. Biswas, R.: Decoding the progress of decision making process in the human cognition systems while evaluating the membership value $\mu(x)$. Issues Intuitionistic Fuzzy Sets Generalized Nets **10**, 21–53 (2013)
29. Bureva, V., Sotirova, E., Atanassov, K.: New operations over intuitionistic fuzzy index matrices. Notes Intuitionistic Fuzzy Sets **18**(4), 12–18 (2012)
30. Chen, S.M.: Cognitive-map-based decision analysis based on NPN logics. Fuzzy Sets Systems **71**(2), 155–163 (1995)
31. Christofides, N.: Graph Theory. Academic Press, New Youk (1975)
32. De, S.K.A., Biswas, A., Roy, E.:. An application of intuitionistic fuzzy sets in medical diagnosis. Fuzzy Sets Syst. **117**(2), 209–213 (2001)
33. Despi, I., Song, G., Chakrabarty, K.: A new intuitionitic fuzzy cognitive maps building method. In: Ding, Y., Li, Y., Fan, Z., Li, S., Wang, L. (eds.) Proceedings of the IEEE International Conference on Fuzzy Systems and Knowledge Discovery, vol 1, pp. 593–597. (2011)
34. Hadjistoykov, P., Atanassov, K.: Remark on intuitionistic fuzzy cognitive maps. Notes Intuitionistic Fuzzy Sets **19**(1), 1–6 (2013)
35. Hadjistoykov, P., Atanassov, K.: Remark on intuitionistic fuzzy cognitive maps with descriptors. Issues Intuitionistic Fuzzy Sets Generalized Nets **10**, 81–88 (2013)
36. Hadjistoykov, P., Atanassov, K.: On temporal intuitionistic fuzzy cognitive maps. Comptes rendus de l'Academie Bulgare des Sciences (in press)
37. Hagan, M., Demuth, H., Beale, M.: Neural Network Design. PWS Publishing, Boston (1996)
38. Haykin, S.: Neural Networks: A Comprehensive Foundation. Macmillan, New York (1994)
39. Iakovidis, D.K., Papageorgiou E.I.: Intuitionistic fuzzy reasoning with cognitive maps. Proceeding of the IEEE International Conference on Fuzzy Systems, pp. 821–827. Taipei, Taiwan, (2011) Accessed 27–30, June 2011
40. Kosko, B.: Fuzzy cognitive maps. Int. J. Man Mach. Stud. **24**(1), 65–75 (1986)
41. Lankaster, P.: Theory of Matrices. Academic Press, New York (1969)
42. Mink, H.: Permanents. Addison-Wesley, Reading (1978)
43. Nagell, T.: Introduction to Number Theory. Wiley, New York (1950)
44. Papageorgiou, E.I., Iakovidis, D.K.: Intuitionistic fuzzy cognitive maps. IEEE Trans. Fuzzy Syst. **21**(2), 342–354 (2013)
45. Parvathi, R., Thilagavathi, S., Thamizhendhi, G. Karunambigai, M.G.: Index matrix representation of intuitionistic fuzzy graphs. Notes Intuitionistic Fuzzy Sets **20**(2), 100–108 (2014)

Index

© Springer International Publishing Switzerland 2014
K.T. Atanassov, *Index Matrices: Towards an Augmented Matrix Calculus*,
Studies in Computational Intelligence 573, DOI 10.1007/978-3-319-10945-9

Printed in the United States
By Bookmasters